Lecture Notes in Computer Science 9557

Commenced Publication in 1973
Founding and Former Series Editors:
Gerhard Goos, Juris Hartmanis, and Jan van Leeuwen

Editorial Board

More information about this series at http://www.springer.com/series/7409

Valentina Tamma · Mauro Dragoni
Rafael Gonçalves · Agnieszka Ławrynowicz (Eds.)

Ontology Engineering

12th International Experiences and Directions Workshop
on OWL, OWLED 2015, co-located with ISWC 2015
Bethlehem, PA, USA, October 9–10, 2015
Revised Selected Papers

Editors
Valentina Tamma
University of Liverpool
Liverpool
UK

Mauro Dragoni
Fondazione Bruno Kessler
Povo, Trento
Italy

Rafael Gonçalves
BMIR, Stanford Center for Biomedical
  Informatics Research
Stanford, CA
USA

Agnieszka Ławrynowicz
Poznan University of Technology
Poznan
Poland

ISSN 0302-9743   ISSN 1611-3349 (electronic)
Lecture Notes in Computer Science
ISBN 978-3-319-33244-4   ISBN 978-3-319-33245-1 (eBook)
DOI 10.1007/978-3-319-33245-1

Library of Congress Control Number: 2016937518

LNCS Sublibrary: SL3 – Information Systems and Applications, incl. Internet/Web, and HCI

Printed on acid-free paper

This Springer imprint is published by Springer Nature
The registered company is Springer International Publishing AG Switzerland

Preface

The OWL: Experiences and Directions Workshop series is an international forum for the OWL community, where practitioners in industry and academia, tool developers and others interested in making use of OWL present research advances, real and potential applications, share experiences, and discuss requirements for language extensions/modifications. OWLED 2015 was the 12th edition of this workshop and was held during October 9–10, in Bethlehem, Pennsylvania, USA, co-located with the International Semantic Web Conference (ISWC 2015).

The technical program featured 20 presentations of accepted full and short papers and two invited talks:

– James Hendler (Rensselaer Polytechnic Institute (RPI)): "On Beyond OWL: Real-World Challenges for Ontologies on the Web"
– Bijan Parsia (University of Manchester): "Two Challenges for OWL"

This year, for the first time we decided to include in the program an *ontology track*, presenting ontologies that pose interesting modelling problems or that can generate challenging tasks with respect to OWLED topics (e.g., ontologies that are challenging for reasoners to handle). There were 30 paper submissions to the workshop, which were reviewed by at least three Program Committee members. Reviews were aimed at constructive feedback and inclusiveness, in order to foster and strengthen the community spirit that characterizes OWLED. Twelve submissions were accepted as full paper with presentation, and three short papers were accepted as short paper with presentation, while five papers were accepted in the ontology track. Among the 20 accepted papers, 18 were accepted for publication in the proceedings volume. We thank the Program Committee for their hard work in reviewing the submitted papers and for the useful feedback they gave to the authors. We would also like to thank the authors for submitting their papers and responding to the reviewers' comments in the final version. We further wish to thank the invited speakers for their inspiring talks. Our thanks also go to Lehigh University, the local organizers of the 14th International Semantic Web Conference for helping us with the logistic organization of OWLED 2015, and the ISWC Organizing Committee. Finally, we would like to thank the development team of the EasyChair conference management system.

October 2015

Valentina Tamma
Mauro Dragoni
Rafael Gonçalves
Agnieszka Ławrynowicz

Organization

Executive Committee

General Chair

Valentina Tamma   University of Liverpool, UK

Program Chairs

Mauro Dragoni	Fondazione Bruno Kessler, Italy
Rafael Gonçalves	Stanford University, USA
Agnieszka Ławrynowicz	Poznan University of Technology, Poland

OWLED Steering Committee

Melanie Courtot	BCCRC, Canada
Matthew Horridge	Stanford University, USA
Pavel Klinov	University of Ulm, Germany
Simon Jupp	EBI, UK
Mariano Rodriguez-Muro	IBM, USA
Bijan Parsia	University of Manchester, UK
Valentina Tamma	University of Liverpool, UK

Program Committee

Valentina Tamma	University of Liverpool, UK
Mauro Dragoni	Fondazione Bruno Kessler, Italy
Agnieszka Ławrynowicz	Poznan University of Technology, Poland
Rafael Gonçalves	Stanford University, USA
Michael Gruninger	University of Toronto, Canada
Marco Rospocher	Fondazione Bruno Kessler, Italy
Krzysztof Janowicz	University of California, Santa Barbara, USA
Vojtěch Svátek	University of Economics, Prague, Czech Republic
Monika Solanki	University of Oxford, UK
Robert Stevens	University of Manchester, UK
Michel Dumontier	Stanford University, USA
Aldo Gangemi	Université Paris 13, France; ISTC-CNR, Italy
Silvio Peroni	University of Bologna and ISTC-CNR, Italy
Rinke Hoekstra	University of Amsterdam/VU University Amsterdam, The Netherlands

Contents

General Terminology Induction in OWL

Viachaslau Sazonau[✉], Uli Sattler, and Gavin Brown

University of Manchester, Oxford Road, Manchester M13 9PL, UK
{sazonauv,sattler,gbrown}@cs.manchester.ac.uk

1 Introduction

An ontology is a machine-processable representation of knowledge about a domain of interest. Ontologies are encoded in formal languages, such as the Web Ontology Language [8], OWL, underpinned by expressive Description Logics, DLs [1]. OWL ontologies are widely-used to represent and share knowledge in application areas such as medicine, biology, astronomy, defence and others.[1] An ontology can contain data and background knowledge (terminology) where both may be incomplete. One might benefit from finding informative correlations in their data taking background knowledge into account. Those correlations may suggest new axioms for the background knowledge or start new inquiries about the data.

However, the problem of terminology induction is generally hard. Firstly, an ideal solution should represent a coherent, self-contained, expert-level modelling. Due to high expressivity of OWL and its Open World Assumption (OWA), the search space can be vast or even infinite depending on the language chosen. Secondly, as usual, the quality of the result depends on the quality of the data which can be incorrect, noisy or insufficient. Ideally, new knowledge should respect the existing knowledge along with the data in order to be maximally informative and avoid contradictions.

Thus, some restrictions and assumptions that simplify the problem are necessary. Another consequence is that any induced knowledge is hypothetical only and requires a domain expert judgement. The contributions of this paper[2] are as follows.

- We state the problem of general terminology induction, i.e. learning sets, called *hypotheses*, of general class inclusions, GCIs, from data (ABox) and background knowledge (TBox).
- We view the problem as multi-objective and define quality criteria for a hypothesis: readability, logical quality, and statistical quality. We define hypothesis quality measures that respect the OWA, interactions between axioms in the hypothesis, and interaction of the hypothesis with the background knowledge.
- We have designed and implemented methods to compute the quality measures.

---

[1] http://bioportal.bioontology.org/.
[2] This is an abridged version of [14].

© Springer International Publishing Switzerland 2016
V. Tamma et al. (Eds.): OWLED 2015, LNCS 9557, pp. 1–13, 2016.
DOI: 10.1007/978-3-319-33245-1_1

– We have designed, implemented and evaluated an anytime algorithm for general terminology induction. We have gained insights into the structure of the search space and developed heuristics to find out promising hypotheses. The experiments show that we can indeed learn interesting hypotheses.

2   Preliminaries

We assume the reader to be familiar with DLs [1] and OWL [8]. The following nomenclature is used throughout this paper. $\mathcal{O} = \mathcal{T} \cup \mathcal{A}$ is an ontology where $\mathcal{T}, \mathcal{A}$ are TBox and ABox, respectively. N_C, N_R, N_I are disjoint and countably infinite sets of class, property, and individual names, respectively. Σ is a signature, $\widetilde{\mathcal{T}}, \widetilde{\mathcal{A}}, \widetilde{\mathcal{O}}$ are signatures of $\mathcal{T}, \mathcal{A}, \mathcal{O}$, respectively. $ind(\mathcal{O}) = N_I \cap \widetilde{\mathcal{O}}$ is a set of individual names occurring in $\mathcal{O}$. α is a general class inclusion, GCI, also called *axiom*. A, B, X, Y are atomic classes (class names), C, D are complex classes (class expressions), R is a property, a, b, c, d are individuals. $mod(\mathcal{O}, \Sigma)$ is a module [7] of an ontology $\mathcal{O}$ given a signature Σ. $\mathbb{C}$ is a set of (possibly complex) classes. H is a hypothesis, $\mathbb{H}$ is a set of hypotheses. In the following, ABox and TBox are called *data* and *background knowledge*, respectively.

3   Related Work

Ontology learning approaches can be characterised along several dimensions. The first one is a type of the data source, e.g. texts, RDF(S), an oracle (a domain expert), positive and negative examples for a class along with the ABox. The second one is a type of the output knowledge, e.g. class descriptions, class inclusions, and its expressivity. The third dimension is methods used: natural language processing, machine learning, association rules mining, oracle queries, Formal Concept Analysis (FCA), least common subsumer (LCS) computation, etc. The fourth dimension is semantics used that can differ from the OWL semantics, e.g. the Closed World Assumption (CWA). One more characteristic is appreciation of available background knowledge. Finally, the degree of domain expert involvement in the learning process greatly varies across approaches. A survey can be found in [12].

We concentrate on learning from instance-level data, i.e. both class and property assertions. Among the approaches aimed at this type of input data are class description learning, CDL [3,5,11], knowledge base completion, KBC [2], association rules mining, ARM [17].

The main method of CDL is machine learning, in particular, Inductive Logic Programming, ILP [13]. The goal is to find a "good" description (class expression) of a given class name from a set of positive and negative examples [11] for it, i.e. learning is *supervised*. The class description must cover all positive and none of the negative examples. Learning is essentially a search in the space of class expressions guided by refinement operators and heuristics. The background knowledge can be used to optimize the search by exploiting the classification hierarchy. To supervise learning, a domain expert has to provide additional information in form of positive and negative examples for a given class, which can be difficult. As a consequence, there are techniques to sample examples from data. In particular, instances of the class are taken as its positive examples and the CWA is made to obtain its negative examples. However, this way can cause problems [10]. Another method of CDL is finding the least common subsumer (LCS) [3]. LCS is computed from the most specific class (MSC) of each instance of a target class. The method, however, is only applicable to weakly expressive languages.

KBC is based on Formal Concept Analysis (FCA) [6]. It is aimed at acquiring (in some sense) complete knowledge bases, in contrast to CDL. KBC requires to define a set of class expressions in advance which can be hard. The degree of domain expert involvement is high as the expert judges axioms and has to supply a counterexample in the case of rejection. One more limitation is that standard FCA can only be applied under the CWA and the OWA of OWL requires modifications of FCA [2].

ARM is yet another approach to ontology learning [17]. Association rules are mined from transaction tables where columns are predefined class expressions which, similarly to the case of KBC, can be difficult to define in advance. In contrast to KBC, ARM, however, permits acquiring axioms that have counterexamples. In contrast to CDL, ARM induces class inclusions and demands neither positive nor negative examples. The approach focuses on weakly expressive languages. Among other restrictions are its CWA and little appreciation of interaction between induced axioms and the background knowledge, as well as mutual interactions between induced axioms, since they are acquired independently.

Thus, ontology learning approaches simplify the problem in different aspects. As a result, there is no approach that has all following capabilities: learns sets of GCIs, appreciates interactions between axioms within the set and interactions of the set with the background knowledge, uses standard OWL semantics, requires no supervision, does not demand frequent human interventions.

4   Settings and Assumptions

This paper is aimed at addressing the problem of inducing general terminological knowledge from data and background knowledge. New knowledge is acquired in the form of *hypotheses*. A hypothesis is a set of axioms which does not contradict the input ontology, i.e. is *consistent* with it, and carries new information, i.e. is *informative* for it.

Definition 1 *(Hypothesis). An axiom α is* informative *for an ontology $\mathcal{O}$ if $\mathcal{O} \not\models \alpha$. A set H of axioms (GCIs) is called a* hypothesis *for an ontology $\mathcal{O}$ if H is consistent with $\mathcal{O}$, i.e. $\mathcal{O} \cup H \not\models \top \sqsubseteq \bot$, and each $\alpha \in H$ is informative for $\mathcal{O}$.*

A hypothesis is evaluated by *quality criteria*: *readability*, *statistical quality*, and *logical quality*. Clearly, a hypothesis can be better on one criterion and worse on another. Therefore, we view terminology induction as a multi-objective problem where objectives are *quality measures* corresponding to the quality criteria. Hypotheses are presented to a domain expert who accepts some of them and rejects others. In order to suggest, or *recommend*, good hypotheses first, a preference relation based on quality measures is imposed on the set of hypotheses. In this paper, we apply the following *settings*.

(i) We use OWL and its standard semantics.
  (a) We allow for the usual OWA, i.e. for an instance a and a class C it is possible that $\mathcal{O} \not\models C(a)$ and $\mathcal{O} \not\models (\neg C)(a)$. As a consequence, data can be regarded as just "incomplete".
  (b) Data normally consists of both class and property assertions, e.g. people with family relations, proteins with interactions between them.
  (c) We consider any logic for which subsumption, $\mathcal{O} \models C \sqsubseteq D$, and instance checking, $\mathcal{O} \models C(a)$, are decidable. We use OWL ontologies and reasoners.
(ii) Any input ontology $\mathcal{O}$ is consistent, i.e. data contains no noise which causes inconsistency.
(iii) Learning is *unsupervised*, i.e. no additional information is required in the form of positive or negative examples.
(iv) A set $\mathbb{C}$ of target (possibly complex) classes is fixed and finite.

The goal of induction is finding good hypotheses over classes $\mathbb{C}$, or $\mathbb{C}$-*hypotheses*. In the following, we only consider $\mathbb{C}$-hypotheses and omit $\mathbb{C}$ from the name. We also define $\mathbb{C}^- := \mathbb{C} \cup \{\neg C \mid C \in \mathbb{C}\}$.

Definition 2 *($\mathbb{C}$-Hypothesis). Given an ontology $\mathcal{O}$, a hypothesis H for $\mathcal{O}$ is called a $\mathbb{C}$-hypothesis if $\alpha \in H$ implies $\alpha = C \sqsubseteq D$, where $C, D \in \mathbb{C}^-$.*

It makes sense to establish a correspondence, sufficient for the task at hand, between an ontology $\mathcal{O}$ and classes $\mathbb{C}$, which we call *projection*.

Definition 3 *(Projection). A projection π of an ontology $\mathcal{O}$ to $\mathbb{C}$ is*

$$\pi(\mathcal{O}, \mathbb{C}) := \{D(a) \mid \mathcal{O} \models D(a) \wedge D \in \mathbb{C}^- \wedge a \in ind(\mathcal{O})\}.$$

Thus, a projection is a set of positive and negative class assertions over classes $\mathbb{C}$ entailed by $\mathcal{O}$. A projection can be viewed as a table where rows are labelled with individuals $ind(\mathcal{O})$ and columns are labelled with classes $\mathbb{C}$. Each cell with indices a, C can contain one of three possible values: "1" if $\mathcal{O} \models C(a)$, "0" if $\mathcal{O} \models \neg C(a)$, "?" if $\mathcal{O} \not\models C(a)$ and $\mathcal{O} \not\models \neg C(a)$. Although there are similarities with a transaction table of ARM, our table view is imaginary only and it permits question marks. We will use the table view for better presentation of examples, see Example 1 and Table 1.

Example 1. Given $\mathbb{C} = \{A, B, \exists R.B\}$, $\mathcal{T} = \varnothing$,

$$\mathcal{A} = \{A(a_1), A(a_2), A(a_3), A(a_4), (\neg A)(b), (\neg A)(c), B(c)$$
$$R(a_1, b), R(a_2, b), R(a_3, b), R(a_4, c)\}.$$

We use the projection to evaluate how well a hypothesis fits the *known* data assuming it is correct on the *unknown* data. Indeed, due to the OWA, a hypothesis can make *assumptions* on the unknown data by turning question marks into ones or zeros. If a hypothesis makes too many assumptions, it may be too "strong", e.g. $H = \{\top \sqsubseteq \sqcap_{C \in \mathbb{C}} C\}$. Therefore, it is necessary to evaluate how "brave" a hypothesis is.

Table 1. Projection for Example 1

	A	B	$\exists R.B$
a_1	1	?	?
a_2	1	?	?
a_3	1	?	?
a_4	1	?	1
b	0	?	?
c	0	1	?

Definition 4 *(Assumption).* An assumption *of a hypothesis H in an ontology $\mathcal{O}$ given $\mathbb{C}$ is*

$$\psi(H, \mathcal{O}, \mathbb{C}) := \{D(a) \mid \mathcal{O} \not\models D(a) \wedge \mathcal{O} \cup H \models D(a) \wedge D \in \mathbb{C}^- \wedge a \in ind(\mathcal{O})\}.$$

As a consequence, $\psi(H, \mathcal{O}, \mathbb{C}) \cap \pi(\mathcal{O}, \mathbb{C}) = \varnothing$ for any hypothesis H. Requiring $\mathcal{O} \not\models (\neg D)(a)$ in Definition 4 is not necessary because if $\mathcal{O} \models (\neg D)(a)$ then H is not a hypothesis due its inconsistency with $\mathcal{O}$. Hypotheses making fewer assumptions are preferred according to Occam's razor.

One can think of suggesting hypotheses as single axioms. However, this approach ignores interactions between axioms that can influence the quality of the hypothesis. Two axioms, which are logically "good" individually, do not necessarily create a logically "good" hypothesis. For example, a hypothesis can become *redundant*, e.g. $H = \{A \sqsubseteq B, \neg B \sqsubseteq \neg A\}$, see Sect. 5.2. In fact, a set of two logically "good" axioms is not necessarily a hypothesis. For example, given that $\{A \sqsubseteq B\}$ and $\{B \sqsubseteq C\}$ are hypotheses for $\mathcal{O}$, a set $\{A \sqsubseteq B, B \sqsubseteq C\}$ is not a hypothesis for $\mathcal{O}$ if $\mathcal{O} \models (A \sqcap \neg C)(a)$. Similar to logical quality, two axioms which are statistically "good" individually may not create a "good" hypothesis.

5 Quality Criteria and Measures for a Hypothesis

5.1 Syntactic Length as a Readability Measure

Readability is the ease with which a hypothesis can be read and understood by a human. One of possible measures of readability is the usual *syntactic length* of a hypothesis.

Definition 5 *(Syntactic Length).* Let A, C, D be (possibly complex) classes, $A \in N_C$ a class name, $R \in N_R$ a property name, $a \in N_I$ an individual name. The syntactic length of a GCI *is defined as follows:* $|C \sqsubseteq D| := |C| + |D|$, where $|\top| = |\bot| = |A| := 1$, $|\neg C| := 1 + |C|$, $|C \sqcap D| = |C \sqcup D| := 1 + |C| + |D|$, $|\exists R.C| = |\forall R.C| := 1 + |C|$, $|\geq nR.C| = |\leq nR.C| := 1 + n + |C|$. The syntactic length of a hypothesis H is $|H| := \sum_{\alpha \in H} |\alpha|$.

5.2 Logical Quality

Logical quality evaluates logical properties of a hypothesis: *logical strength* and *redundancy*. Logical strength is commonly called generality in machine learning.

Definition 6 *(Logical Strength). A hypothesis H is* weaker *(more general) than another hypothesis H' if $H' \models H$ and $H \not\models H'$.*

A hypothesis can contain axioms which are superfluous, or *redundant*, within the hypothesis, even if those axioms are informative. For example, axiom $A \sqsubseteq C$ is redundant in hypothesis $\{A \sqsubseteq B, B \sqsubseteq C, A \sqsubseteq C\}$ and axiom $\neg B \sqsubseteq \neg A$ is redundant in hypothesis $\{A \sqsubseteq B, \neg B \sqsubseteq \neg A\}$. Axioms can also have *redundant parts*. For example, D is a redundant part of axiom $A \sqsubseteq C \sqcap D$ in hypothesis $\{A \sqsubseteq B \sqcap D, A \sqsubseteq C \sqcap D\}$.

Definition 7 *(Redundancy). A hypothesis H is* redundant *if there exists a hypothesis H' such that $H' \equiv H$ and $|H'| < |H|$. Otherwise, H is* non-redundant.

Lemma 1. *If a hypothesis H is non-redundant, then $|H| = min\{|H'| \mid H' \equiv H\}$.*

We define the logical strength and redundancy of a hypothesis H regardless of $\mathcal{O}$. The reason is that an axiom $\alpha \in H$, which is informative for $\mathcal{O}$ and non-redundant in H, can be interesting, even if it is not informative for $\mathcal{O} \cup H\backslash\{\alpha\}$. Such axiom reveals yet only implicit (and possibly unknown) relation between classes. Additionally, the search for good hypotheses would require entailment checking $\mathcal{O} \cup H \models H'$ which could make it infeasible for hard ontologies.

5.3   Statistical Quality

Statistical quality criteria are aimed at selecting hypotheses that best represent data given background knowledge. In order to comply with the standard OWL semantics and its OWA, we consider the statistical quality of a hypothesis as twofold. Firstly, hypotheses fit data and background knowledge differently. Secondly, hypotheses make different number of assumptions in data given background knowledge, i.e. some hypotheses are more cautious than others.

Fitness and Braveness. In order to evaluate the statistical quality of a hypothesis, we exploit the idea that axioms can encode regularities in the data. Those regularities can be used to "compress" the data, i.e. to present it in a shorter way. This is the fundamental principle of the *minimum description length induction* [4,16]. According to it, the better a hypothesis fits the data, the shorter description of the data it provides.

A standard way of measuring the description length is using syntactic measures. However, syntactic measures do not respect logical interactions of a hypothesis with data and background knowledge. Therefore, we introduce a semantic measure of the description length. We define fitness and braveness of a hypothesis as follows.

Definition 8 *(Description Length, Fitness, Braveness). The description length of an ABox $\mathcal{B}$ given an ontology $\mathcal{O} = \mathcal{T} \cup \mathcal{A}$ is*

$$minSize(\mathcal{B}, \mathcal{O}) := min\{|\mathcal{B}'| \mid \mathcal{B}' \cup \mathcal{O} \equiv \mathcal{B} \cup \mathcal{O}\}.$$

Given an ontology $\mathcal{O}$, a set $\mathbb{C}$ of classes, and a hypothesis H, let $\pi := \pi(\mathcal{O}, \mathbb{C})$ and $\psi := \psi(H, \mathcal{O}, \mathbb{C})$. Then

(i) fitness *of H is* $fit(H, \mathcal{O}, \mathbb{C}) := |\pi| - minSize(\pi, \mathcal{T} \cup H)$,
(ii) braveness *of H is* $bra(H, \mathcal{O}, \mathbb{C}) := minSize(\psi, \mathcal{O})$.

As a consequence of Definition 8, all semantically equivalent hypotheses have the same fitness and the same braveness which is stated by Lemma 2.

Lemma 2. *Given an ontology $\mathcal{O}$, a set $\mathbb{C}$ of classes, and two hypotheses H_1, H_2, if $H_1 \equiv H_2$ then $fit(H_1, \mathcal{O}, \mathbb{C}) = fit(H_2, \mathcal{O}, \mathbb{C})$ and $bra(H_1, \mathcal{O}, \mathbb{C}) = bra(H_2, \mathcal{O}, \mathbb{C})$.*

Fitness of a hypothesis indicates how well the projection can be shrunk using the hypothesis and background knowledge, i.e. a better shrinkage corresponds to a better fitness. Braveness of a hypothesis measures how many assumptions it makes in the data given the background knowledge.

As a consequence of Definition 8, fitness and braveness are semantically sound and syntax independent measures of the statistical quality of a hypothesis. They take into account both the interaction of a hypothesis with the background knowledge and interactions between axioms within the hypothesis. The measures respect the standard OWL semantics, in particular, they deal with its OWA and, consequently, with incomplete data. Finally, they demand no supervision, such as positive or negative examples, and no additional information besides the input ontology.

6   General Terminology Induction

According to Definition 1, we only consider hypotheses which are logically sound, i.e. informative and consistent with the background knowledge and data. The goal of the induction is finding among those hypotheses ones which have maximal fitness and minimal braveness, or better represent the data.

We impose a readability constraint on a hypothesis: it must not exceed a given syntactic length. The logical weakness of a hypothesis is reflected by its braveness: weaker hypotheses have a lower braveness and are preferred (respecting their fitness) according to Occam's razor. A redundant hypothesis has the same fitness and braveness as its non-redundant counterpart but a greater length that might be occupied by better axioms. We state the problem of general terminology induction in OWL as follows.

Definition 9 *(General Terminology Induction). Given an ontology $\mathcal{O}$ and a set $\mathbb{C}$ of classes, the problem of* general terminology induction *is to find all best hypotheses which do not exceed length ℓ.*

Thus, as in ILP, we view induction as search in the space of hypotheses restricted by a language bias, determined by $\mathbb{C}$ and ℓ in our case. We regard the process of constructing hypotheses as being equivalent to ranking them in a justified way which is based on fitness and braveness.

7   Implementation and Evaluation

7.1   Implementation

Tools and Hardware. All algorithms are implemented in Java 7 using OWL API (3.5.0). We use the OWL 2 DL reasoner FaCT++ (1.6.3) [15] which supports incremental reasoning. The experiments are executed on the following machine: Linux Ubuntu 14.04.2 LTS (64 bit), Intel Core i5-3470 3.20 GHz, 8 GB RAM.

7.2   Evaluation

Evaluation Goals. The experiments are aimed at answering the following questions.

Q1 Where are we likely to find good hypotheses: in more expressive languages for $\mathbb{C}$ or bigger values of ℓ?

Q2 How does expressivity of the language and maximal length of a hypothesis influence the performance of computing the fitness and braveness?

Q3 Can we acquire hypotheses that seem plausible, so that we can use them to enrich our background knowledge, or that tell us interesting information about our data?

Choice of Ontologies. We conduct the empirical evaluation on a corpus of ontologies selected from related work [5,10] including DL-Learner datasets,[3] Protégé OWL,[4] and TONES[5] repositories. The Kinship ontology is obtained from UCI Machine Learning Repository.[6] We have selected the ontologies based on the following criteria. Firstly, data contains both class and property assertions, at least 15 individuals. Secondly, ontology classification takes less 10 min. Thirdly, we are sufficiently confident that we understand the topic of the ontology. The corpus is available online.[7]

Table 2 describes the corpus where we use the following metrics. $|ind(\mathcal{A})|$, CA, RA are numbers of individuals, concept and property assertions in the ABox, respectively. $degree(\mathcal{A})$, $conn(\mathcal{A})$ are the average degree and average number of individuals in a connected component, respectively. $|\widetilde{\mathcal{A}}|$, $|\widetilde{\mathcal{T}}|$ are sizes of the ABox and TBox signature. $Jac(\widetilde{\mathcal{A}}, \widetilde{\mathcal{T}})$ is the Jaccard index[8] of ABox and TBox signatures, $open(\mathcal{A}, \mathcal{T})$ is the average number of question marks per individual-class name pair.

---

[3] https://github.com/AKSW/DL-Learner.

[4] http://protegewiki.stanford.edu/index.php/Protege_Ontology_Library.

[5] http://owl.cs.manchester.ac.uk/repository/.

[6] https://archive.ics.uci.edu/ml/datasets/Kinship.

[7] http://www.cs.man.ac.uk/~sazonauv/tbox_induction/corpus/.

[8] The size of the intersection divided by the size of the union.

Table 2. Ontologies and their metrics

	DL	$|ind(A)|$	CA	RA	$degree(A)$	$conn(A)$	$|\tilde{A}|$	$|\tilde{T}|$	$Jac(\tilde{A},\tilde{T})$	$open(A,T)$
Alzheimer	$\mathcal{AL}$	150	106	854	5.7	150	40	0	0	0.96
Arch	$\mathcal{ALC}$	19	26	26	1.4	3.8	10	13	0.77	0.53
BasicFamily	$\mathcal{ALI}$	31	50	95	3.1	10.3	6	6	1	0.67
Carcinogenesis	$\mathcal{ALC(D)}$	22372	22372	40666	1.8	65.8	113	146	0.77	0.65
Cinema	$\mathcal{ALCOF}$	45	45	76	1.7	45	7	37	0.19	0.88
Earthrealm	$\mathcal{SHOIN(D)}$	171	179	203	1.2	7.4	23	2482	0.01	0.89
Economy	$\mathcal{ALCH(D)}$	482	649	555	1.2	5.3	29	380	0.04	0.94
Financial	$\mathcal{ALCOIF}$	17941	17941	47248	2.6	8970.5	52	76	0.68	0.54
GeoSkills	$\mathcal{ALCHOIN(D)}$	2592	4681	3896	1.5	13.9	569	618	0.90	0.69
Heart	$\mathcal{AL(D)}$	280	275	1080	3.9	280	9	11	0.82	0.90
Kinship	$\mathcal{ALI}$	24	116	40	1.7	12	18	4	0.16	0.81
KRK	$\mathcal{SHI}$	420	525	1508	3.6	4	25	40	0.55	0.65
Mammographic	$\mathcal{AL(D)}$	975	975	2883	3.0	975	18	22	0.82	0.97
MDM073	$\mathcal{ALCHOF(D)}$	112	130	169	1.5	2.0	82	215	0.38	0.51
Mutagenesis	$\mathcal{AL(D)}$	14145	14145	26533	1.9	61.5	60	91	0.66	0.99
NTN	$\mathcal{SHOIN(D)}$	724	724	1636	2.3	2.8	64	78	0.82	0.96
Suramin	$\mathcal{AL(D)}$	2979	2979	6008	2.0	175.2	20	49	0.41	0.97

Evaluation Setup. To answer the raised questions, we set up the following experimental pipeline. We choose maximal length ℓ from $\{2, 4, 6, 8, 10\}$. In order to generate classes $\mathbb{C}$, we use signature $\Sigma := \widetilde{M}$, where $M = \top\bot\text{-}module(T, \tilde{A})$. We investigate 5 class languages G_i, such that $G_i \subseteq G_{i+1}$ (duplicates are avoided by the means of OWL's structural equivalence):

$G_1 := \{X \mid X \in \Sigma\}$;
$G_2 := G_1 \cup \{X_M \mid X_M \text{ is a possibly complex subclass in } M\}$;
$G_3 := G_2 \cup \{X \sqcap Y \mid X, Y \in \Sigma\}$;
$G_4 := G_3 \cup \{\exists R.X \mid X, R \in \Sigma\}$;
$G_5 := G_4 \cup \{X \sqcap \exists R.Y \mid X, Y, R \in \Sigma\}$.

Given an ontology $\mathcal{O}$, for each combination of a class language G and maximal length ℓ we run the algorithm (see [14] for details) with 10 min timeout. Once the algorithm terminates, we record the fitness and braveness of each hypothesis and the average hypothesis evaluation time which comprises computing its fitness and braveness. Finally, we store up to 100 best hypotheses.

7.3   Results

The dependence of fitness and braveness on language and length is shown on Fig. 1. The values obtained are normalised, i.e. divided by the maximal value.

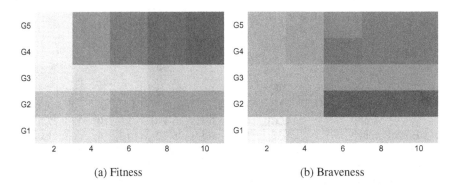

(a) Fitness   (b) Braveness

Fig. 1. Dependence of fitness (a) and braveness (b) on language expressivity and maximal length: darker colours reflect greater numbers

Then, the values are aggregated across the corpus and the average value is reported per cell.

Our first observation is that some languages and lengths result in no hypotheses induced which happens if a class language is not expressive enough or hypothesis length is too low. We aggregate and average only over non-empty values. An expected observation is that increasing expressivity is useless if an ontology is poor, e.g. contains few relations in the data and axioms in the background knowledge. On the other hand, if an ontology is rich, increasing expressivity may or may not be fruitful.

Figure 1 shows that increasing length always results in hypotheses of higher fitness and mostly, but not always, of higher braveness since added axioms may make no assumptions or repeat the assumptions already made. Increasing expressivity also generally leads to higher fitness and higher braveness. However, the changes are not as gradual as for length, in particular, braveness seems irregular. Interestingly, we observe that G_2 consistently outperforms G_3 in fitness, despite $G_2 \subseteq G_3$, which can be explained as follows. On the one hand, the search space considerably increases from G_2 to G_3. On the other hand, G_3 appears to be less fruitful than G_2 (compare to G_4 and G_5). As a result, it becomes harder to find equally good hypotheses in the same time. Thus, the answer to Q2 is that increasing expressivity and length promises better fitness but commonly worse braveness.

We also observe that the average hypothesis evaluation time does not vary widely. Thus, the answer to Q2 is that performance does not degrade significantly for higher expressivity and length. The performance of evaluating a hypothesis is as follows: less than 0.1 s for 8 ontologies, from 0.1 to 1 s for 4 ontologies, from 1 to 10 s for 4 ontologies, and around 15 s for 1 ontology. The results can be found online.[9]

In order to answer Q3, we act as domain experts and eyeball the induced hypotheses. We aim at finding plausible and interesting hypotheses. Some results

---
[9] http://www.cs.man.ac.uk/~sazonauv/tbox_induction/results/.

Table 3. Examples of hypotheses induced within 10 min

Ontology	Examples of hypotheses
Alzheimer	$Drug \sqsubseteq \exists getsReplacedBy.Substituent$
	$Substituent \sqsubseteq \exists hasPolatisation.Polar$
	$\exists hasPolatisation.Polar \sqsubseteq \exists isHAcceptor.HAcceptor$
Arch	$construction \sqsubseteq \exists hasPillar.pillar$
	$\exists hasParallelpipe.wedge \sqsubseteq \exists hasPillar.freeStandingPillar$
	$\exists touches.pillar \sqsubseteq \exists leftof.pillar$
BasicFamily	$\exists hasChild.Person \sqsubseteq Person$
	$\exists hasParent.Person \sqsubseteq Person$
	$\exists hasParent.Female \sqsubseteq \exists hasParent.Male$
Cinema	$Movie \sqsubseteq \exists hasForActor.Actor$
	$Movie \sqsubseteq \exists hasForGenre.Genre$
	$\exists hasForActor.\{Eastwood\} \sqsubseteq \exists hasForGenre.\{Western\}$
	$\exists hasForDirector.\{Burton\} \sqsubseteq \exists hasForActor.\{Depp\}$
Earthrealm	$\exists hasDefaultUnit.BaseUnit \sqsubseteq \exists hasDefaultUnit.ComplexUnit$
	$\exists hasDefaultUnit.\{second\} \sqsubseteq TimeRelatedQuantity$
	$\exists hasDefaultUnit.\{meterPerSecond\} \sqsubseteq DrySeasonDuration$
Economy	$Nation \equiv IndependentState$
	$\exists economyType.EconomicDevelopmentLevel$
	$\sqsubseteq \exists economyType.IMFDevelopmentLevel$
Financial	$Account \sqsubseteq \exists hasStatementIssuanceFrequency.Monthly$
	$\exists isOwnerOf.Account \sqsubseteq Client$
Mammographic	$\exists hasMargin.spiculated \sqsubseteq \exists hasShape.irregular$
	$\exists hasShape.irregular \sqsubseteq \exists hasDensity.low$
Mutagenesis	$Compound \sqsubseteq \exists hasBond.Bond1$
	$\exists inBond.Hydrogen3 \sqsubseteq Bond1$
	$\exists inBond.Oxygen40 \sqsubseteq \exists inBond.Nitrogen38$
NTN	$Man \equiv \forall spouseOf.Woman$
	$\exists knows.Man \sqsubseteq Man$
	$\exists relativeOf.Man \sqsubseteq Man$

are shown in Table 3. Firstly, we observe that induced hypotheses can, in fact, enrich the background knowledge, see Table 3. If the background knowledge is poor, as in **BasicFamily** and **Cinema**, or even absent, as in **Alzheimer**, hypotheses seem to be a good starting point for modellers. If the background knowledge is incomplete, hypotheses appear to be interesting missing bits, e.g. for **Economy**, **Financial**, **NTN**, and **Mutagenesis**.

Secondly, we observe that hypotheses can reveal interesting relations in our data. This can expose new knowledge about the domain and help to understand the data. For example, hypotheses discover relations between particular actors, directors, and movie genres from **Cinema**. Another example is **Mammographic** where we can learn relations between diagnostic observations, e.g. having irregular shape implies having lower density. Such hypotheses can potentially inform doctors of yet unknown relations in their data, facilitate future research in the domain, and lead to data improvements, e.g. a supplement of images of tumours that have irregular shape and high density.

Thirdly, hypotheses can contain "strange" axioms which may help us highlight, on the one hand, odd or erroneous modelling and, on the other hand, inaccurate or abnormal data. We observe this for **Arch** inducing $\exists touches.pillar \sqsubseteq$

$\exists leftof.pillar$ (why is there nothing to the right?) and for **Earthrealm** inducing $\exists hasDefaultUnit.\{meterPerSecond\} \sqsubseteq DrySeasonDuration$ (wrong unit?). Thus, we can answer Q3 positively.

Although we use different settings and the goal of induction is different, we make some comparison of our results with related work. In particular, we consider the supervised CDL and its implementation DL-Learner [11]. Given a set of positive and negative examples for a target class *construction* in **Arch**, it searches for definition $construction \equiv \exists hasPillar.(freeStandingPillar \sqcap \exists leftof.\exists supports.\top)$. As Table 3 shows, our approach induces a weaker definition of *construction* along with some related knowledge. For **Cinema** we observe that descriptions of different movie types are induced, e.g. $EastwoodMovie \sqsubseteq \exists hasForActor.\{Eastwood\}$, $EastwoodMovie \sqsubseteq \exists hasForGenre.\{Western\}$. For **NTN** the definition $Man \equiv \forall spouseOf.Woman$ is induced. Thus, although our approach is unsupervised, it shows the potential to learn class definitions.

8 Discussion and Future Work

The evaluation shows that our approach is able to induce interesting hypotheses. On the one hand, they can potentially be helpful to build and improve the background knowledge. On the other hand, hypotheses seemingly discover new knowledge about the domain and help us understand the data. Interestingly, they may help us identify modelling errors and data flaws.

Although the search space is vast, general terminology induction is feasible. It is encouraging given that statistically and logically sound measures are used to evaluate a hypothesis and this requires reasoning. We observe that larger and more expressive hypotheses are generally better and still feasible.

As for future work, we will investigate more informed ways of constructing a set of promising initial classes, e.g. using techniques from CDL, along with new algorithms and heuristics for search space exploration. We will also attempt to extend the methodology to deal with noisy data that causes inconsistency, e.g. using techniques from [9]. We plan to investigate learning property hierarchies.

We intend to go beyond the corpus and carry out case studies with domain experts to evaluate our approach in more detail. We also consider other scenarios, e.g. how acceptance or rejection of a hypothesis affects other hypotheses, how hypotheses can be used for predicting class memberships of individuals, terminology abduction and "what if" analysis of data under the OWA.

References

1. Baader, F., Calvanese, D., McGuinness, D., Nardi, D., Patel-Schneider, P.F. (eds.): The Description Logic Handbook: Theory, Implementation, and Applications. Cambridge University Press, Cambridge (2003)
2. Baader, F., Ganter, B., Sertkaya, B., Sattler, U.: Completing description logic knowledge bases using formal concept analysis. In: Proceedings of the 20th International Joint Conference on Artifical Intelligence, IJCAI 2007, pp. 230–235. Morgan Kaufmann Publishers Inc., San Francisco (2007)

3. Baader, F., Sertkaya, B., Turhan, A.Y.: Computing the least common subsumer w.r.t. a background terminology. J. Appl. Logic **5**(3), 392–420 (2007)
4. Conklin, D., Witten, I.H.: Complexity-based induction. Mach. Learn. **16**(3), 203–225 (1994)
5. Fanizzi, N., d'Amato, C., Esposito, F.: DL-FOIL Concept Learning in Description Logics. In: Železný, F., Lavrač, N. (eds.) ILP 2008. LNCS (LNAI), vol. 5194, pp. 107–121. Springer, Heidelberg (2008)
6. Ganter, B., Wille, R.: Formal Concept Analysis, vol. 284. Springer, Berlin (1999)
7. Grau, B.C., Horrocks, I., Kazakov, Y., Sattler, U.: Modular reuse of ontologies: theory and practice. J. Artif. Intell. Res. **31**, 273–318 (2008)
8. Grau, B.C., Horrocks, I., Motik, B., Parsia, B., Patel-Schneider, P., Sattler, U.: OWL 2: the next step for OWL. Web Semant. **6**(4), 309–322 (2008)
9. Haase, P., Stojanovic, L.: Consistent evolution of OWL ontologies. In: Gómez-Pérez, A., Euzenat, J. (eds.) ESWC 2005. LNCS, vol. 3532, pp. 182–197. Springer, Heidelberg (2005)
10. Lehmann, J., Auer, S., Bühmann, L., Tramp, S.: Class expression learning for ontology engineering. Web Semant. **9**(1), 71–81 (2011)
11. Lehmann, J., Hitzler, P.: Concept learning in description logics using refinement operators. Mach. Learn. **78**(1–2), 203–250 (2010)
12. Lehmann, J., Völker, J. (eds.): Perspectives On Ontology Learning, Studies in the Semantic Web, vol. 18. IOS Press, Amsterdam (2014)
13. Muggleton, S.: Inductive logic programming. New Gener. Comput. **8**(4), 295–318 (1991)
14. Sazonau, V., Sattler, U., Brown, G.: General terminology induction in OWL. In: Arenas, M., et al. (eds.) ISWC 2015. LNCS, vol. 9366, pp. 533–550. Springer, Heidelberg (2015)
15. Tsarkov, D., Horrocks, I.: FACT++ description logic reasoner: system description. In: Furbach, U., Shankar, N. (eds.) IJCAR 2006. LNCS (LNAI), vol. 4130, pp. 292–297. Springer, Heidelberg (2006)
16. Vitányi, P.M., Li, M.: Minimum description length induction, bayesianism, and kolmogorov complexity. IEEE Trans. Inf. Theor. **46**(2), 446–464 (2000)
17. Völker, J., Niepert, M.: Statistical Schema Induction. In: Antoniou, G., Grobelnik, M., Simperl, E., Parsia, B., Plexousakis, D., De Leenheer, P., Pan, J. (eds.) ESWC 2011, Part I. LNCS, vol. 6643, pp. 124–138. Springer, Heidelberg (2011)

OBOWLMorph: Starting Ontology Development from PURO Background Models

Marek Dudáš[⊠], Tomáš Hanzal, Vojtěch Svátek, and Ondřej Zamazal

Department of Information and Knowledge Engineering, University of Economics,
W. Churchill Sq.4, 130 67 Prague 3, Czech Republic
{marek.dudas,tomas.hanzal,svatek,ondrej.zamazal}@vse.cz

Abstract. We propose adding two additional steps to OWL ontology development and offer tools supporting it. A so-called PURO background model of an example situation to be covered by the ontology is first created, then a seed of the ontology is generated automatically from it, allowing users to choose suitable modeling style and import the ontology seed into a common ontology editor where it can be finalized. Using PURO as intermediary model should enable better collaboration, documentation and early detection of design problems. The paper focuses on OBOWLMorph: a tool for ontology generation from a PURO model.

1 Introduction

In the semantic web realms, the prevailing practice of formalizing ontologies is writing them, from the onset, in OWL, merely starting from textual specifications and informal charts. The advantages of OWL as uniform representation of ontologies throughout all 'formal' phases of their development lifecycle are its thorough standardization, solid support by authoring tools, and powerful reasoning abilities allowing formal consistency checking of the models. On the other hand, the direct transition from informal specifications to OWL puts quite high demands on ontology engineers. In software engineering, UML models are often created before the actual coding. Database designers use even two levels of intermediate models: conceptual and logical. The common benefits of such intermediate models are better collaboration, documentation and early detection of architectural or logical problems. Ontology engineers directly defining OWL entities based on informal specifications have to deal with two problems at the same time: (a) "What are the entities and relations inherently described in the specification?" and (b) "How to represent them with OWL constructs?" Moreover, the latter question often has several possible answers – choosing different *OWL modeling styles*. Therefore we investigate, and attempt to offer tools for, a more stepwise approach to ontological engineering where a 'seed' of the ontology is first drafted and exemplified in a less constrained language (called PURO [10]) and a basis of the ontology is then generated from it through pattern-based PURO-to-OWL transformation. This allows focusing on question (a) in the first step without having to solve question (b), which is dealt with in the second step.

© Springer International Publishing Switzerland 2016
V. Tamma et al. (Eds.): OWLED 2015, LNCS 9557, pp. 14–20, 2016.
DOI: 10.1007/978-3-319-33245-1_2

The generated basis of an ontology can then be finalized using common ontology editors. While such a practice is not new per se (since some earlier methodologies [6] proposed to create the first formalization in first-order predicate calculus), the crucial point is the use of PURO as language with structure similar to OWL, giving way to the automatic transformation. Simple PURO-to-OWL transformation has recently been integrated [4] as additional feature into PURO Modeler (our graphical PURO model authoring tool prototype); it allows to display alternative OWL representations of uncomplicated PURO models, thus serving as an auxiliary tool for ontology developers, with educative role for novices. In this paper, in contrast, we present customizable PURO-to-OWL transformation functionality implemented in a dedicated tool, OBOWLMorph, which has the potential to play a more central role in the ontology development workflow.

2 PURO Language and OWL Modeling Styles

OWL ontology developer can often choose different combinations of language constructs to model the same situation. The choice might be driven by the intended usage of the ontology: web markup vocabularies often favor 'feature' assignment to entities through data properties, while linked data vocabularies prefer object properties for this purpose; reasoning-enabled ontologies, in turn, express 'features' as classes. We call sets of such choices *modeling styles*.[1]

PURO is an ontological modeling language recently drafted as common interlingua for different modeling styles in OWL. A model built in PURO is denoted as *ontological background model* (OBM). For example, the fact that a concrete book is a 'paperback' can be expressed, in OWL, using an object property assertion, a data property assertion or a class instantiation. In PURO it is always the last option, called 'B-instantiation' ('background-instantiation'), since 'in the background', individual paperback books and the notion of 'paperback' are interrelated via sound set membership.

PURO inventory is very similar to that of OWL, assuring easy understandability by OWL-bred engineers. It is based on two distinctions: between particulars and universals and between relationships and objects (hence the PURO acronym). There are six basic entity types: B-object (particular object), B-type (type of object/type), B-relationship (particular relationship), B-relation (type of relationship), B-valuation (particular assertion of quantitative value) and B-attribute (type of valuation). An OBM consists of named entities of these types, plus of subTypeOf and instanceOf relationships. Obviously, the 'object-type-relation' triad of PURO corresponds to the 'individual-class-property' triad of OWL, except that 1) PURO does not limit the arity of relations and allows higher-order classes, and, consequently, 2) enables abstracting from modeling style choices that are enforced by these limitations in OWL. For example the fact that a book is published by a publisher in a certain year normally requires reification to a new entity (e.g., of 'PublishingEvent' type) in OWL; in PURO a ternary relationship suffices for this purpose. Every PURO OBM describes a concrete sample situation. *Instances* ('particulars') play central role in the model,

---

[1] In agreement with [3] where OWL feature modeling styles are analysed.

helping designers to avoid speculating about abstract categories for which there would be no concrete data available and making them focus on sample situations to be covered. Particulars also glue different type-level entities ('universals') together in a contiguous model. A PURO OBM thus does not only map on OWL ontologies but also on samples of their respective fact bases (Aboxes).

The research of modeling styles is still in its infancy. To test OBOWLMorph, we implemented five ad-hoc modeling styles. In *Data property style*, data properties are used whenever possible. *Object property style* prefers object properties, even to model subTypeOf relationships and B-attributes. *Object-prop and subclassing style* is similar to *Object property style*, but subTypeOf is represented by *rdfs:subClassOf* and B-attributes are modeled as data properties. In *Reification style*, even binary B-relationships are reified into classes and pairs of object properties, otherwise it is same as *Object-prop and subclassing style*. So is *Class membership style* with the exception that binary B-relationships are turned into classes of subjects of the B-relationship having the value of the object.

3 Related Research

The most similar to our approach is OntoUML [2]: a conceptual modeling language based on UML and grounded in Universal Foundational Ontology (UFO). OLED, the graphical editor for OntoUML, allows to transform it into OWL fragments. The transformation is hard-coded and each OntoUML element has its single OWL counterpart. The users can however select for each OntoUML element whether it may change in time or is 'read-only', and the choice is reflected in the transformation; such functionality is planned to be added to our framework as well. Bauman [1] implemented XSLT transformation of conceptual models into XML Schema, while OWL as target is only mentioned as possible future work. The user can choose a sort of modeling style, e.g., whether to transform a concept to an XML attribute or child-element. To allow reusing existing ER diagrams, Fahad [5] designed their rule-based transformation to OWL ontologies. The framework is however not intended as a general ontology development alternative. For transformation between different types of models such as UML, XML Schema or OWL, Kensche et al. [8] suggested to employ a generic metamodel (GeRoMe), as an abstraction of particular metamodels. In order to uniformly capture specific properties of models of different types, elements of GeRoMe are decorated with a set of role objects (e.g., a role attribute is mapped to a column in a relational schema and to a data property in OWL DL). Native models can be imported/exported into/from GeRoMe. In all mentioned OWL generation methods, the input model is created at the level of types. In our approach, in contrast, the input model is created as an example situation at the instance level. Finally, since the PURO language has also been proposed as means of formally testing the conceptual coherence of ontologies [10], it can be compared to OntoClean [7]; it differs in the 'meta-properties' assigned to entities.

4   OBOWLMorph Implementation and Example of Usage

The generation of OWL from an OBM is done with SPARQL. To allow that, the OBM is first serialized into RDF using a simple 'PURO vocabulary'. The serialization also includes information about the desired modeling style (gathered from the user) in the form of annotations of serialized PURO entities. The serialized OBM is then transformed to OWL with a set of SPARQL UPDATE[2] queries. Using SPARQL allows the transformation rules to be easily altered and extended by the semantic web community. So far we have defined 12 SPARQL patterns,[3] covering the most common combinations of PURO entities and their OWL representations in different modeling styles. The WHERE part of each query represents a pattern of an OBM fragment, including the modeling style annotations. The INSERT part describes a corresponding OWL fragment. The resulting OWL fragment is inserted into a separate RDF graph. All SPARQL queries are applied automatically in a sequence and the resulting OWL ontology 'seed' is then extracted from the graph. OBOWLMorph is implemented as a web application[4] and integrated with PURO Modeler, a visual OBM editor connected to same DB as OBOWLMorph.

The OBOWLMorph interface consists of two windows: one displays the loaded OBM, while the other shows the OWL ontology seed generated from the OBM and visualized in WebVOWL [9]. The user can choose a different target OWL modeling style for each OBM entity: s/he simply clicks on an entity, selects from available modeling styles shown in a pop-up window and clicks the 'update' button to see the change in the OWL ontology seed. A default modeling style is used for the entities unaffected by the user.

Use-case Scenario Example: Consider that during the development of a food ontology, a verbal example is gathered: *A boiled egg is a dish of size 100 g, containing 12 g of fat, 2 g of carbohydrates and 800 kJ of energy. Its ingredient is one egg.* The knowledge engineer creates its OBM in PURO Modeler[5] as shown in Fig. 1. S/he may share and discuss it with other developers and check whether all concepts from the example are modeled and correctly labeled. Then s/he proceeds to OWL modeling. Keeping the default modeling style settings (*Object-prop and subclassing style* set for all entities), OBOWLMorph produces the result shown in Fig. 2. The engineer now considers the intended usage of the ontology and decides to, e.g., simplify it by changing the modeling style of the "egg" entity to *Datatype*. On the other hand, the "energy" value needs to be modeled as instance, because the engineer decided to allow adding the unit specification to avoid confusion between kJ and calories. Therefore, s/he sets the *Reification* modeling style for the "800 kJ" entity. After updating, the ontology seed looks as shown in Fig. 3. When the engineer is satisfied with the result, s/he may download the OWL ontology seed, import it to an ontology editor such as Protégé, and continue working on it.

---

[2] Used instead of CONSTRUCT for implementation-specific reasons.

[3] Available at http://lod2-dev.vse.cz/puromodeler-v2/OBOWLMorph/patterns/.

[4] http://lod2-dev.vse.cz/puromodeler-v2/OBOWLMorph/.

[5] Following the guidelines available at http://bit.ly/1MFr8Lm (in development).

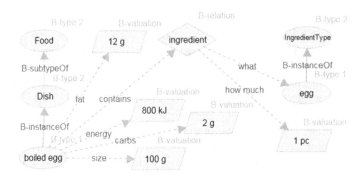

Fig. 1. OBM of a boiled egg as a dish with one ingredient and nutrition info.

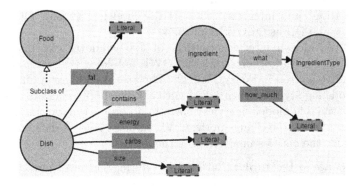

Fig. 2. OWL ontology seed generated from the OBM using default modeling style.

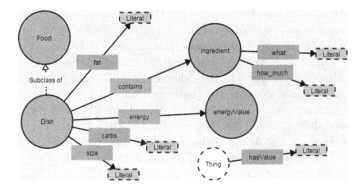

Fig. 3. OWL ontology seed with modeling style on some OBM entities altered.

5   Conclusions and Future Work

We experimentally implemented an application introducing two more steps into ontology design, analogical to logical or UML models in DB and SW engineering: the ontology designer first creates a PURO ontological background model, from which a seed of the desired OWL ontology is generated automatically. Desired OWL modeling style for each part of the background model can be selected.

The current inventory of modeling styles is a proof-of-concept one; however, further styles can be implemented rapidly, being mere SPARQL UPDATE queries. Future work will also include reuse of entities (with fitting style) from existing vocabularies in addition to coining new ones, during transformation, as well as exploitation of entity naming conventions when using sophisticated patterns, e.g., those including reification. Thorough evaluation by user assessment and comparison to gold-standard ontologies is also foreseen.

The research is supported by UEP IGA F4/90/2015, by the H2020 project no. 645833 (OpenBudgets.eu) and by long-term institutional support of research activities by Faculty of Informatics and Statistics, Univ. of Economics, Prague. Ondřej Zamazal is supported by CSF 14-14076P.

References

1. Bauman, B.T.: Prying apart semantics and implementation: generating xml schemata directly from ontologically sound conceptual models. In: Proceedings of Balisage: The Markup Conference 2009. http://www.balisage.net/Proceedings/vol3/print/Bauman01/BalisageVol3-Bauman01.html
2. Benevides, A.B., Guizzardi, G.: A model-based tool for conceptual modeling and domain ontology engineering in OntoUML. In: Filipe, J., Cordeiro, J. (eds.) Enterprise Information Systems. LNBIP, vol. 24, pp. 528–538. Springer, Heidelberg (2009)
3. Dermeval, D., Tenório, T., Bittencourt, I.I., Silva, A., Isotani, S., Ribeiro, M.: Ontology-based feature modeling: an empirical study in changing scenarios. Expert Syst. Appl. **42**(11), 4950–4964 (2015)
4. Dudáš, M., Hanzal, T., Svátek, V., Zamazal, O.: OBM2OWL patterns: spotlight on OWL modeling versatility. In: Workshop on Ontology and Semantic WebPatterns (WOP) at ISWC (2015). http://lod2-dev.vse.cz/puromodeler/purom-wop15.pdf
5. Fahad, M.: Er2owl: generating OWL ontology from ER diagram. In: Shi, Z., Mercier-Laurent, E., Leake, D. (eds.) Intelligent Information Processing IV. IFIP, vol. 288, pp. 28–37. Springer, Heidelberg (2008)
6. Gómez-Pérez, A., Fernández-López, M., Corcho, O.: Ontological Engineering. Springer, London (2004)
7. Guarino, N., Welty, C.A.: An overview of OntoClean. In: Staab, S., Studer, R. (eds.) Handbook on ontologies. International Handbooks on Information Systems, pp. 201–220. Springer, Heidelberg (2009)
8. Kensche, D., Quix, C., Chatti, M.A., Jarke, M.: *GeRoMe*: a generic role based metamodel for model management. In: Spaccapietra, S., Atzeni, P., Fages, F., Hacid, M.-S., Kifer, M., Mylopoulos, J., Pernici, B., Shvaiko, P., Trujillo, J., Zaihrayeu, I. (eds.) Journal on Data Semantics VIII. LNCS, vol. 4380, pp. 82–117. Springer, Heidelberg (2007)

9. Lohmann, S., Negru, S., Haag, F., Ertl, T.: VOWL 2: user-oriented visualization of ontologies. In: Janowicz, K., Schlobach, S., Lambrix, P., Hyvönen, E. (eds.) EKAW 2014. LNCS, vol. 8876, pp. 266–281. Springer, Heidelberg (2014)
10. Svátek, V., Homola, M., Kluka, J., Vacura, M.: Metamodeling-based coherence checking of OWL vocabulary background models. In: OWLED (2013)

A Similarity Based Approach to Omission Finding in Ontologies

Tahani Alsubait[(⊠)], Bijan Parsia, and Uli Sattler

School of Computer Science, The University of Manchester, Manchester, UK
{alsubait,bparsia,sattler}@cs.man.ac.uk

Abstract. With the growing interest in using ontologies in semantically-enabled applications, the interest in enhancing the quality of such ontologies has grown as well. Standard reasoning services focus on certain obvious dimensions of quality, e.g., to detect inconsistencies and incoherence. In addition, bespoke tools have been presented to address the completeness dimension of quality, e.g., missing entailments. These tools are usually focused on very restricted subsets of all the possible missing entailments, i.e., only atomic subsumptions. We present a new protocol to detect both existing invalid entailments and missing valid entailments. We also present a case study to evaluate the usefulness of the presented protocol for ontology validation purposes.

1 Introduction

With the growing interest in using ontologies in semantically-enabled applications, the interest in enhancing the quality of such ontologies has grown as well. Ontologies can grow large in terms of size and complexity, making it challenging to maintain their quality and accuracy. Typically, the ontology development life cycle involves an ontology validation stage in which both ontology developers and domain experts come together to review the ontology and make sure it is free of errors. The most hard-to-spot errors in ontologies are the ones that do not make the ontology inconsistent or incoherent, though cause either undesirable or missing entailments. This is similar to the so-called "logical errors" in programming languages which cause the program to produce undesired output but do not cause compilation errors or abnormal termination. However, as is the case with programming languages, standard debugging tools cannot help in identifying such logical errors. This is why there is a need to develop tools and techniques for this purpose.

Indeed, there are many possible ways to find errors in ontologies. Direct ontology inspection can be effective but has the obvious disadvantage of being infeasible for large ontologies. In addition, direct inspection might be more effective for finding soundness problems (i.e., invalid entailments) rather than completeness problems (i.e., missing entailments) [9]. Other approaches have been proposed to address completeness problems. For example, Formal Concept Analysis (FCA) has been used for such a purpose [5]. FCA, in this context, is used to compute

© Springer International Publishing Switzerland 2016
V. Tamma et al. (Eds.): OWLED 2015, LNCS 9557, pp. 21–32, 2016.
DOI: 10.1007/978-3-319-33245-1_3

a concept lattice, i.e. a subsumption hierarchy of all conjunctions of concept names occurring in an ontology, and the negations of these concept names. This lattice is then used to present successive questions to a domain expert to identify possible missing terminological or assertional axioms. A general observation about this approach is that they focus on finding missing atomic subsumptions. That is, they address *only* the completeness dimension (i.e., ignore soundness) and within the completeness dimension, they *only* consider subsumption relations between concept names (i.e., ignore complex subsumptions). Similarly, the approach presented by Lambrix et al. [8] is aimed at completing is_a hierarchies.

In this paper, we define a new protocol for finding omissions in ontologies. These omissions can be either missing *atomic subsumptions* or missing *complex subsumptions*. The protocol invloves asking a domain expert a set of multiple choice questions (MCQs) with high similarity between the correct and wrong answers.[1] Restricting the answer set to only those answers that are very similar to the correct answer can be useful in restricting the search space (in a principled way) when attempting to detect omissions. Using similarity to elicit knowledge from domain experts has already been used in well known elicitation techniques. For example, the triadic elicitation technique involves presenting 3 concepts to domain experts who are asked to identify the two similar concepts and explain why the third is considered different. Similarly, we present some statements that are entailed to be invalid by the ontology, yet they are very similar to a valid entailment and ask the experts to verify whether they are indeed invalid entailments or possibly missing valid entailments.

The questions presented to the expert should take the form of a multiple-response question[2] where the expert is asked to select all (and only) the correct answers. We re-use the question generation (QG) application described in [2] to generate questions that have exactly one answer entailed by the ontology to be correct. For the purpose of using these questions to validate the ontology, we select (for each question) a varied number, ranging for example from 1 to 10, of answers that are entailed to be wrong answers. The similarity between the key and distractors is set to be above a threshold. To measure the similarity between (possibly) complex concepts, we use the similarity measures presented in [3,4]. To examine whether using a high value threshold has an impact on the number and type of the errors identified, we experiment with two different thresholds as we will describe in detail in Sect. 3. In general, since the wrong answers are selected to be similar to the correct answer, we question whether the ontology should entail that they are correct answers as well, i.e., a missing entailment.

2   Motivation

As an example, consider the Java ontology that has been used in [1] as a knowledge source for generating educationally-useful MCQs. A detailed description of

---

[1] In MCQ terminology, a correct answer is referred to as a key and a wrong answer is referred to as a distractor.

[2] In a multiple-response question, more than one answer can be correct.

the ontology is presented in [1]. During the development of the Java ontology, we have witnessed the usefulness of looking at MCQs generated from this ontology for validating it on the fly. Some important "errors" in the ontology were easily identified by looking at the MCQs generated from it, in particular, MCQs with errors. Some errors were syntactic (e.g., typing mistakes) while others were logical (e.g., a wrong entailment identified by looking at an invalid key or a missing entailment identified by looking at an invalid distractor). Logical errors are generally harder to spot and considered more interesting when debugging an ontology. We briefly present some specific examples from the Java ontology in Tables 1 and 2.

Table 1. Missing entailment example

Stem:	A feature of Virtual Machine Code is:
Key:	(A) Portability
Distractors:	(B) Write once Run Anywhere
	(C) Platform Independence
	(D) Reusability
Explanation of error:	Wrong answers are invalid (i.e., they are features of Virtual Machine Code)
Reasons for the (missing) entailment:	Portability $\sqsubseteq \exists$ isAFeatureOf.VirtualMachineCode
	WriteOnceRunAnywhere $\sqsubseteq \exists$ isAFeatureOf.JavaProgramming
	PlatformIndependence $\sqsubseteq \exists$ isAFeatureOf.JavaProgramming
	Reusability $\sqsubseteq \exists$ isAFeatureOf.JavaProgramming
	All the above classes (features) have the common subsumer $\exists$ isAFeatureOf.Top (hence, are similar; therefore, they all appeared in the answer list of this MCQ)

Table 2. Undesired entailment example

Stem:	Swing stands for:
Key:	(A) Application Programming Interface
Distractors:	(B) Abstract Windowing Toolkit
	(C) Java Foundation Classes
Explanation of error:	the key is not a correct answer (i.e., Swing does not stand for Application Programming Interface)
Reasons for the (undesired) entailment:	Swing $\sqsubseteq$ API
	API $\sqsubseteq \exists$ standsFor.ApplicationProgramming Interface
	Therefore, the ontology entails that:
	Swing $\sqsubseteq \exists$ standsFor.ApplicationProgramming Interface

Clearly, some logical errors in the Java ontology have resulted in producing the errors that appear in these MCQs. Identifying the errors in these MCQs by a Java expert has helped in finding and correcting some omissions in the Java ontology. These examples show that looking at questions generated from an ontology can be fruitful for identifying some omissions in the ontology. In particular, it helped to identify invalid keys and distractors, i.e., answers that were thought to be correct while they are in fact wrong or vice versa.

In this paper, we present a case study to further explore the applicability of QG methods for ontology validation purposes. Rather than validating an ontology under development, we study the case of validating a previously built ontology in an attempt to suggest ways to improve it. We present some specific examples for possible errors in the SNOMED CT ontology as identified by some domain experts. In addition, QG methods can support ontology comprehension purposes which can be a goal in itself or it can be done prior to validating an ontology that has been built by a different ontology developer. We briefly tackle this in the study presented in this paper.

3   Implementing a Prototype QG-Based Application for Ontology Validation

To evaluate the usefulness of the suggested QG-based approach for ontology validation purposes, we have implemented a prototype web-based application that (1) presents a selected set of multiple-response questions generated from an ontology to a domain expert (see Fig. 1) and (2) based on the expert's answers, the application suggests some possible wrong and/or missing entailments in the ontology (see Fig. 2). As we already described in the introduction, the questions are generated such that they have only one answer which is entailed by the ontology to be correct. However, experts answering these questions are asked to pick all the answers they believe to be correct. Experts are also asked to indicate whether they are confident about their answers, per question. They can also leave a comment for a detailed explanation.

When the answers provided by an expert are different from the ones entailed by the ontology, the expert is asked to confirm their answers, as shown in Fig. 3. The aim of this extra verification step is to encourage rethinking about the answer.

4   A Case Study

4.1   Goals

The main goal of this case study is to evaluate the usefulness of the suggested QG-based approach for ontology validation purposes. To address this goal, we try to answer the following question: Can a domain expert identify some omissions in an ontology by looking at MCQs generated from that ontology? We focus on a specific class of MCQs in which each wrong answer is similar to the correct

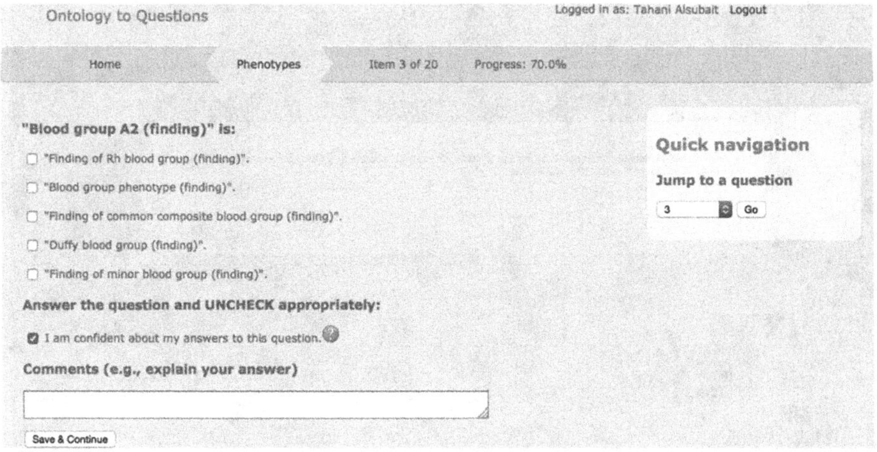

Fig. 1. QG-based support for ontology validation

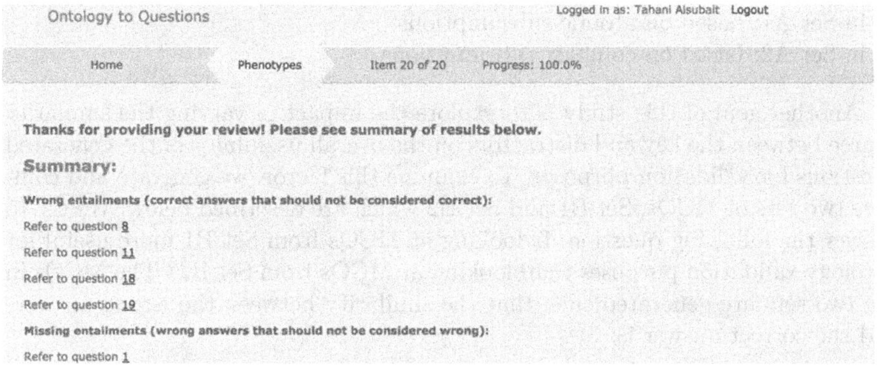

Fig. 2. Summary of suggestions to improve the ontology

answer (but entailed by the ontology to be a wrong answer). We expect that looking at such questions can reveal some omissions or missing statements (in the ontology) that might be difficult to spot without looking at the questions. This is because these wrong answers are similar to the correct answer and therefore raise the question of whether they have been considered as wrong answers due to having any missing statements in the ontology or due to actual constraints in the domain. The missing statements that are intended to be detected can be either *atomic* or *complex* subsumptions. Missing or invalid atomic subsumptions highlight problems in the inferred class hierarchy of the ontology. Since this hierarchy is frequently looked at by ontology developers, we expect, in general, that there are more missing/invalid complex subsumptions rather than atomic subsumptions in a given ontology. We examine this hypothesis in the current study by looking at two sets of questions, Set A1 and Set A2. The questions in the two sets are constructed:

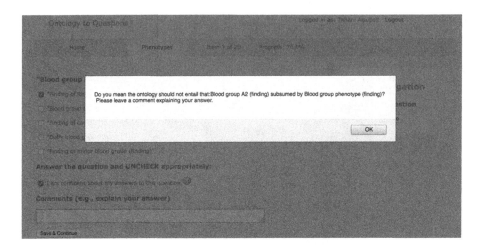

Fig. 3. Extra verification step

1. in Set A1: based on atomic subsumptions.
2. in Set A2: based on complex subsumptions.

Another goal of this study is to explore the impact of varying the similarity degree between the key and distractors on the overall usefulness of the generated questions for validation purposes. To examine this factor, we generate and compare two sets of MCQs, Set B1 and Set B2 which are described below. We try to answer the following question: Is looking at MCQs from Set B1 more useful for ontology validation purposes than looking at MCQs from Set B2? The MCQs in the two sets are generated such that the similarity between the wrong answers and the correct answer is:

1. in Set B1: above a threshold Δ_{max}.
2. in Set B2: below a threshold Δ_{max} but above a second threshold Δ_{min}.

The two sets A1 and A2 are not disjoint from sets B1 and B2. To examine all possibilities, we generate four disjoint sets of questions such that the questions:

1. in Set 1: are selected from Set A1 and Set B1.
2. in Set 2: are selected from Set A1 and Set B2.
3. in Set 3: are selected from Set A2 and Set B1.
4. in Set 4: are selected from Set A2 and Set B2.

4.2   Materials and Methods

Ontology Selection. The current study requires the availability of a domain expert to answer a set of MCQs generated from a domain ontology. Due to the availability of an expert in BioInformatics, we asked that expert to select some parts of an ontology which he thinks might be suitable for the purpose of this study. Due to the expert's interest in SNOMED CT in general and genetic findings in particular and his assumptions that the ontology is not detailed enough in this part, we selected a (small) part of genetic findings that covers phenotypes (e.g., Blood groups). All the subclasses (197 classes) of the class *Phenotype* were used as a seed signature to extract a ⊥-module [10]. In addition, the object property *RoleGroup* was added to the seed signature. This property is used to group certain properties together [12] and is necessary for extracting the module. The resulting module has a total of 246 classes and 6 object properties.

Generating Questions. Two sets of questions were generated from the extracted module using the prototype QG application described in [2]. This prototype generates two different sets of questions, namely difficult and easy questions. The difficult questions are generated such that the similarity between the key and distractors is above the average similarity between all siblings in the ontology (or in the current study, the extracted module). The easy questions are generated such that the similarity between the key and distractors is above two thirds of the average similarity between all siblings in the module (but less than the average similarity between all siblings). For the current study, we consider difficult questions to be questions of Set B1 and easy questions to be questions of Set B2. After computing the average similarity between all siblings in the module, the thresholds Δ_{max} and Δ_{min} have been set to 0.88 and 0.587, respectively. The generated questions take the form "What is X?" where X is a class name and the answers are either class names or class expressions. This kind of questions is suitable for finding missing/invalid entailments that we are interested in. Among the generated questions, 223 questions have class-name-based answers, referred to as Set A1 questions, and 24 questions have class-expression-based answers, referred to as Set A2 questions. Among the class-expression-based questions, only 5 questions are suitable for Set B1 (i.e., the similarity between the key and distractors is above the threshold Δ_{max}). These 5 questions are referred to as Set 3 as defined in Sect. 4.1. Each question has exactly one key but the number of distractors was variable. If the number of generated distractors for a given question is more than 10, we randomly select 10 distractors out of the available ones. We have restricted the number of distractors to be below or equal to 10 to make the question answering phase manageable.

Answering Questions. Two domain experts have been asked to answer a total of 20 questions (5 questions from each of the four sets Set 1, Set 2, Set 3 and Set 4). The first expert is a bioinformatician and the second expert is a physician. The 20 questions were selected randomly from the set of generated questions in the previous step. Three samples of those questions are presented in Sect. 4.3. The questions were presented to the domain experts via the web-interface described in Sect. 2, see Fig. 1. The first 10 questions are from Set A1 and the second 10 questions are from Set A2. We chose to present questions from Set A1 first, to captivate the reader, because they are expected to take less time to answer compared to questions from Set A2. Within Sets A1 and

A2, questions from Sets B1 and B2 are randomly ordered. Also, a think-aloud technique was used to get a deeper insight into the advantages and limitations of the approach. The experts were allowed to use any external source to help them in answering the questions. After answering all the questions, the experts were asked to answer three last questions about their overall experience in answering the questions. These questions are shown in Fig. 4.

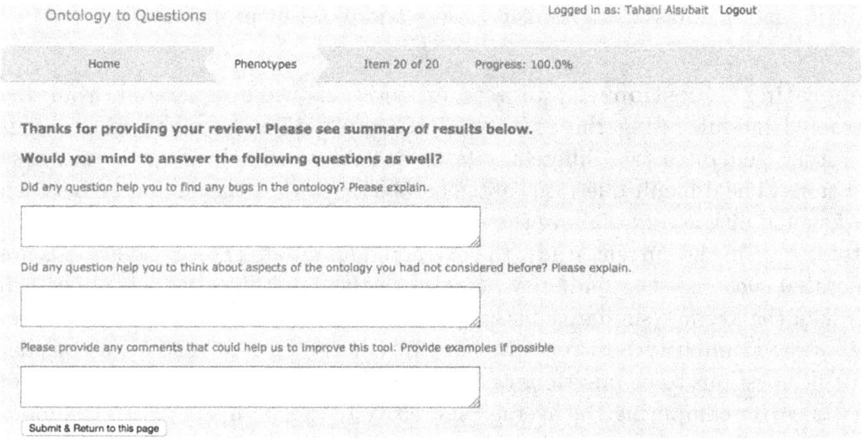

Fig. 4. Using QG-methods to validate ontologies

4.3   Results and Discussion

For 9 out of the 10 questions in Set B1, the first expert's answers were correct, i.e., equivalent to what is entailed by the ontology. The only question for which this expert's answers were different from the ones entailed by the ontology is the question presented in Table 3. This question is the only question which has an answer that contains an existential restriction; all the other answers contain either class names or conjunctions of class names. The expert has identified both a missing entailment (invalid wrong answer) and a wrong entailment (invalid correct answer). In particular, the expert indicated that the ontology should entail that a finding of common composite blood group is subsumed by a finding of blood group and phenotype finding. He also indicated that the ontology should not entail that a finding of common composite blood group is subsumed by a finding of blood group and interprets (attribute) ABO and Rho(D) typing (procedure). The expert indicated that he was not confident about his answers to this question and explained that by reporting that he was not familiar with the terminology used by the ontology to describe the concepts presented in this question, e.g., interprets (attribute). In consistent with the first expert's answers, the second expert answered all the questions in Set B1 correctly; hence she did not identify any possible omissions in this part of the ontology.

For 8 out of the 10 questions in Set B2, the first and second experts' answers were equivalent to what is entailed by the ontology. The two questions for which

Table 3. A first example for a question generated from SNOMED CT

Stem:	"Finding of common composite blood group" is:
Key:	(A) "Finding of blood group" and Interprets "ABO and Rho(D) typing"
Distractors:	(B) "Finding of blood group" and "Phenotype finding"

the two experts' answers were different from the ones entailed by the ontology are the questions presented in Tables 4 and 5. In both questions, the answers are conjunctions of class names. Again, in both questions, the experts have identified a missing entailment (by selecting one of the distractors) and a wrong entailment (by not selecting the expected key). Both experts have agreed on the wrong answer that they chose to select as an answer. The two experts have indicated that they are not confident about their answers to these two questions. The first expert explained why he was not confident about his answers to the question presented in Table 4 by pointing out that one of the terms used in the question, i.e., inherited, seems irrelevant since all blood groups are inherited. For this question, the experts indicated that the ontology should entail that inherited weak D phenotype is subsumed by blood group phenotype and finding of minor blood group. Similarly, for the question presented in Table 5, the experts indicated that the ontology should entail that trans weak D phenotype is subsumed by blood group phenotype and finding of minor blood group.

In total, the first expert indicated that he was confident when answering only 7 questions out of the 20 questions. The second expert was confident when answering 13 questions out of the 20 questions. The first expert explained that by pointing out that although the terminology used in the ontology might seem to be natural to an ontology developer, it does not seem to be natural for a subject matter expert. Consistent with this, the second expert reported that the language of questions made it difficult to interpret what the question was asking. The first expert also reported that the questions seem to be of varying difficulty. For example, he pointed out that answering most of the questions from Set A1 was straightforward. These questions use only class names as answers. In contrast, he reported that two questions from the same set, which also use only class names as answers, were harder to answer. He explained that by pointing out that the answers were very similar and hence he found it difficult to decide which answer is the correct answer. The answers to these questions were: Blood laboratory and Blood bank which are indeed similar (yet refer to different departments). The first expert further explains that he selected what he thought was the best answer, rather than the only correct answer. Consistent with this, the second expert reported that, for the exact two questions, she picked what she thought was the best answer. The experts did not identify any missing entailments in these two questions, i.e., they did not indicate that a wrong answer should be a correct answer. However, their explanation supports the hypothesis we are testing in this study, i.e., looking at MCQs with distractors that are similar to the key can be helpful in identifying missing entailments.

Table 4. A second example for a question generated from SNOMED CT

Stem:	"Inherited weak D phenotype" is:
Key:	(A) "Blood group phenotype" and "Finding of Rh blood group"
Distractors:	(B) "Blood group phenotype" and "Finding of ABO blood group"
	(C) 'Blood group phenotype" and "Duffy blood group"
	(D) "Blood group B" and "Blood group Para-Bombay"
	(E) 'Blood group phenotype" and "Finding of minor blood group"

Table 5. A third example for a question generated from SNOMED CT

Stem:	"Trans weak D phenotype" is:
Key:	(A) "Blood group phenotype" and "Finding of Rh blood group"
Distractors:	(B) "Blood group phenotype" and "Duffy blood group"
	(C) "Blood group B" and "Blood group Para-Bombay"
	(D) "Blood group phenotype" and "Finding of minor blood group"
	(E) "Blood group phenotype" and "Finding of ABO blood group"

As described earlier, the similarity between the key and distractors in questions from Set B1 is higher than the similarity between the key and distractors in questions from Set B2. Although one would expect that questions in Set B1 would reveal more omissions in the ontology compared to questions in Set B2 (because the wrong answers are more similar to the correct answers), this was not the case. Questions in Set B1 have identified 2 (possible) omissions while questions in Set B2 have identified 4 (possible) omissions. This can be explained by the fact that errors can occur in *different* parts of the ontology. For example, questions in Set B1 would identify missing subsumees that are very close to their (potential) subsumer, e.g., in the inferred class hierarchy. In contrast, questions in Set B2 would identify missing subsumees that are not very close to their potential subsumer. In general, looking at this (rather small) set of questions was helpful in spotting some omissions in the ontology and suggesting improvements. Consistent with our expectations, the results also show that the method may be generally more helpful in identifying invalid/missing entailments involving complex subsumptions, i.e., Set A2, rather than atomic subsumptions, i.e., Set A1.

The aim of the second and third question presented to the experts after answering the questions was to evaluate the usefulness of the presented MCQs to support ontology comprehension purposes. According to the answers provided by the experts, the questions were not very helpful in identifying new aspects of the ontology they had not considered before. The first expert pointed out that this is due to having (1) questions that seem to be unnatural to a subject matter expert (due to describing concepts in an uncommon way) and (2) changes in the difficulty level of the questions (partly due to the first point). He further explains by pointing out that these two points might limit the usefulness of this form of MCQs for supporting students who want to learn about the subject. The second expert, who is a physician, did not respond to this question as she was not familiar with the ontology.

4.4 Related Work

Baader et al. [5] presented a FCA-based approach for completing Description Logics-based knowledge bases. Their approach is aimed at extending both the terminological and the assertional part of the knowledge base, i.e., the TBox and the ABox, respectively. A *Protégé* plugin implementing this approach is presented in [11].

Bertolino et al. [6] have investigated the use of QG-based methods for validation purposes. Their method aims at validating models in general and can be applied to ontologies as well. A set of True/False questions generated from an (altered) model are presented to a group of domain experts. The responses gathered from domain experts are used to validate the model. The method proposed by Bertolino et al. is different from our method in that they suggest to alter the model by deliberately introducing some errors in it before the QG step. Their method is also suitable for finding invalid entailments but not missing entailments. Although they have reported that their method have helped the recruited experts to think about new aspects of the domain which they have not considered before, the method does not guarantee that this applies to the unaltered (error-free) parts of the domain only.

Another related work is the approach presented by Dragisic et al. [7] that takes already found missing entailments as input and suggest logical solutions to repair the ontology by possibly adding missing axioms. Their approach is extended in [8] by attempting to repair ontologies without given missing entailments. This approach is different from our approach in that it only considers missing atomic subsumptions.

5 Summary and Future Directions

We have suggested a new protocol for finding omissions in OWL ontologies. We have also presented a case study for evaluating the usefulness of the suggested protocol for ontology validation purposes. Although the results seem to be promising, they are far from significant. Further efforts are needed to improve and evaluate the presented strategy. In particular, more user studies are needed. As a future work, we plan to implement a *Protégé* plugin to allow ontology developers to benefit from the suggested protocol.

References

1. Alsubait, T., Parsia, B., Sattler, U.: Generating multiple choice questions from ontologies: How Far Can We Go? In: Lambrix, P., Hyvönen, E., Blomqvist, E., Presutti, V., Qi, G., Sattler, U., Ding, Y., Ghidini, C. (eds.) EKWA 2014 Satellite Events. LNCS, vol. 8982, pp. 66–79. Springer, Heidelberg (2015)

2. Alsubait, T., Parsia, B., Sattler, U.: Generating multiple choice questions from ontologies: Lessons learnt. In: The 11th OWL: Experiences and Directions Workshop (OWLED 2014) (2014)

3. Alsubait, T., Parsia, B., Sattler, U.: Measuring similarity in ontologies: a new family of measures. In: Janowicz, K., Schlobach, S., Lambrix, P., Hyvönen, E. (eds.) EKAW 2014. LNCS, vol. 8876, pp. 13–25. Springer, Heidelberg (2014)

4. Alsubait, T., Parsia, B., Sattler, U.: Measuring similarity in ontologies: How bad is a cheap measure? In: 27th International Workshop on Description Logics (DL-2014) (2014)

5. Baader, F., Ganter, B., Sertkaya, B., Sattler, U.: Completing description logic knowledge bases using formal concept analysis. In: Proceedings of IJCAI 2007 (2007)

6. Bertolino, A., DeAngelis, G., DiSandro, A., Sabetta, A.: Is my model right? let me ask the expert. J. Syst. Softw. **84**(7), 1089–1099 (2011)

7. Dragisic, Z., Lambrix, P., Wei-Kleiner, F.: Completing the is-a structure of biomedical ontologies. In: Galhardas, H., Rahm, E. (eds.) DILS 2014. LNCS, vol. 8574, pp. 66–80. Springer, Heidelberg (2014)

8. Lambrix, P., Wei-Kleiner, F., Dragisic, Z.: Completing the is-a structure in lightweight ontologies. J. Biomed. Semant. **6**(12) (2015)

9. Rogers, J.: Development of a methodology and an ontological schema for medical terminology. Ph.D. thesis, Department of Computer Science (2004)

10. Sattler, U., Schneider, T., Zakharyaschev, M.: Which kind of module should I extract? In: Proceedings of the 22nd International Workshop on Description Logics (DL-2009) (2009)

11. Sertkaya, B.: ONTOCOMP: A PROTÉGÉ plugin for completing owl ontologies. In: Aroyo, L., et al. (eds.) ESWC 2009. LNCS, vol. 5554, pp. 898–902. Springer, Heidelberg (2009)

12. Spackman, K., Dionne, R., Mays, E., Weis, J.: Role grouping as an extension to the description logic of ontylog, motivated by concept modeling in snomed. In: Proceedings of the AMIA Symposium: American Medical Informatics Association, p. 712 (2002)

An Ontology for Supporting the Evolution of Virtual Reality Scenarios

Mauro Dragoni[1]([✉]), Chiara Ghidini[1], Paolo Busetta[2], Mauro Fruet[2], and Matteo Pedrotti[2]

[1] FBK–IRST, Trento, Italy
{dragoni,ghidini}@fbk.eu
[2] Delta Informatica, Trento, Italy
{paolo.busetta,mauro.fruet,matteo.pedrotti}@deltainformatica.eu

Abstract. Serious games with 3D interfaces are Virtual Reality (VR) systems that are becoming common for the training of military and emergency teams. A platform for the development of serious games should allow the addition of semantics to the virtual environment and the modularization of the artificial intelligence controlling the behaviors of non-playing characters in order to support a productive end-user development environment. In this paper, we report the ontology design activity performed in the context of the **PRESTO** project aiming to realize a conceptual model able to abstract the developers from the graphical and geometrical properties of the entities in the VR, as well as the behavioral models associated to the non-playing characters.

1 Introduction

Serious games with 3D interfaces are a branch of VR systems and are often used for the training of military personnel (in individual as well as team coordination danger situations) and, more recently, for the training of civilian professionals (firefighters, medical personnel, etc.) in emergency situations using tools such as VBS3[1] and XVR[2].

A crucial step towards the adoption of VR for training is the ability to configure scenarios for a specific training session at reduced costs and complexity. By looking at state of the art technologies, it is already possible to do so for physical landscapes, physical phenomena, and crowds (including their behaviors), and trainers and system integrators can assemble and customize serious game products for a specific scenario using commercial products and libraries that need to be (easily) adapted to the specific landscapes and needs of the clients.

Current attempts to the programming of non playing characters rely on ad hoc specifications/implementations of their behaviors done by VR developers. Thus, a specific behavior (e.g., a function emulating a panicking reaction) is hardwired to a specific item (e.g., the element "Caucasian_boy_17" of a VR such

---

[1] https://www.bisimulations.com/.

[2] http://futureshield.com/xvr-esemble.shtml.

V. Tamma et al. (Eds.): OWLED 2015, LNCS 9557, pp. 33–44, 2016.
DOI: 10.1007/978-3-319-33245-1_4

as XVR) directly in the code. This generates a number of problems typical of ad hoc, low level solutions: the solution is scarcely reusable, it often depends on the specific knowledge of the code of a specific developer, and is cumbersome to modify, since every change required by the trainer has to be communicated to the developers and directly implemented in the code in a case by case manner. The existence of high level specifications of non playing characters and modular behaviors, described in a manner that is independent from the specific VR, and available for both trainers and developers, would be an important step towards the definition of reusable, flexible, and therefore cheaper, scenarios that include non playing characters.

In this paper, we focus on the experience of using Semantic Web techniques, and in particular lightweight ontologies, for the high level description of the artificial entities (including characters) and their behaviors in gaming in order to uncouple the description of scenarios performed by the trainers from their physical implementation in charge to the developers. Differently from a number of works in literature that often uses ontologies for a detailed description of the geometrical properties of space and objects, the focus of our work is on the description of the entities of a VR scenario from the cognitive point of views of the trainers and the developers alike, in a way that is semantically well founded and independent of a specific game or scenario [1], and with the goal of fostering clarity, reuse, and mutual understanding [2].

To the best of our knowledge, the construction of the ontology presented in this paper provides a first experience towards the description of a virtual world from a cognitive level that can highlight the potential and criticality of using Semantic Web techniques, and existing ontologies, to describe a VR from a cognitive point of view and can provide the basis for further developments.

2 The PRESTO Project

The objective of PRESTO (Plausible Representation of Emergency Scenarios for Training Operations) research project is the creation of a system for the customization of serious games scenarios based on VRs. The advantage of this system, compared to the state of the art, resides in the richness and the ease of defining the behavior of artificial characters in simulated scenarios, and on the execution engines able to manage cognitive behaviors, actions, and perceptions within a VR environment. One of the main outcome of the project is the possibility of specifying procedures, psychological profiles, and other factors that influence the behavior of individuals and/or small groups in any role (emergency teams, victims, observers, terrorists, criminals, etc.) and to build scenarios, for instance a car accident, in which part or all of the people involved are simulated by artificial characters. To this end, the system has to include an environment for building the training scenarios by the VR trainer, tools for the specification of cognitive and perceptual models used for augmenting psychological profiles of non-player characters, and execution engines able to manage cognitive behaviors, actions, and perceptions within a VR environment.

The system can be used, for example, for training safety personnel, for the verification and the optimization of operational procedures, and for the analysis of work environments. The system has been tested in a pilot use case selected in a specific application domain of large interest in both commercial and research fields: training for emergency management within close environments (such as fires, evacuations, overload of users due to external factors such great disasters scale, etc.). The pilot has been be conducted in collaboration with the Health Services of the Trentino local government (APSS).

The open problems addressed by this project may be summarized as follows:

1. the perception of the virtual environment by an artificial character and the execution of its models and procedures must be able to adapt to the context, to its history and status (fatigue, emotions, intake of stimulants such as caffeine or depressants such as alcohol) and must maintain a level of variability (i.e. in the accuracy of the vision, the rate of reaction, in the choices among alternatives) such that the behavior is plausible but not trivially predictable;
2. the representation of procedures and patterns of behavior must be independent of one specific usage scenario and accessible to training specialists (i.e. industrial safety or civil protection) rather than just a computer, in an environment facilitating the definition and configuration of training scenarios by such specialists.

The first open problem relates to aspects such as the usage of a BDI (Beliefs-Desire-Intention) multi-agent system with cognitive extensions, CoJACK [3], as the artificial intelligent engine for the generation/selection of behaviors in serious games [4], that go beyond the scope of this paper.

What we present in this work, instead, is the experience of using Semantic Web techniques, and in particular lightweight ontologies, to contribute to the second open problem, that is the development of a programming environment for serious game platforms thanks to end-user development tools [5] and the ability to mix and match scenario components (including behavioral components) taken off-the-shelf from a market place.

3 PRESTO Ontology Design

The development of programming environment for the high level description of artificial entities (including characters) and their behaviors in scenarios of serious games requires the ability to represent a wide range of entities that *exist* in the (artificial) world. The approach taken in PRESTO is to use ontologies to represent this knowledge, in a way that is semantically well specified and independent of a specific game or scenario [1].

The construction of the PRESTO ontology therefore is driven by typical questions that arise when building ontological representations of a domain, that is:

– "What are the entities that exist, or can be said to exist, in a Virtual Reality scenario?"

– "How can such entities be grouped, related within a hierarchy, and subdivided according to similarities and differences?"

Differently from Ontology in philosophy, where these questions are motivated from the need to investigate the nature and essence of being, we have looked at these questions from the pragmatic point of view of computer science, where ontologies and taxonomic representations have been widely proposed and used to provide important conceptual modeling tools for a range of technologies, such as database schemas, knowledge-based systems, and semantic lexicons [2] with the aim of fostering clarity, reuse, and mutual understanding.

A serious problem we had to face in PRESTO was the lack-of/limited-availability of training experts and software developers, and the broad scope of items and behaviors that can occur in an arbitrary scenario of VR, that can range from terrorist attacks in a war zone, to a road accidents in a motorway, to a fire alarm in a nuclear plant or hospital and so on. Because of that reason, building everything from the ground up by relying on domain experts and using one of the state of the art ontology engineering methodologies such as METHONTOLOGY [6] was deemed unfeasible. Thus the process followed in PRESTO has been driven by an attempt to: (1) maximize the reuse of already existing knowledge and (2) revise and select this knowledge with the help of experts by means of more traditional ontology engineering approaches such as the one mentioned above. The choice of already existing knowledge has lead us to consider the following two sources:

– state of the art foundational ontologies which provide a first ontological characterization of the entities that exist in the (VR) world; and
– the concrete items (such as people, tools, vehicles, and so on) that come with VR environments and can be used to populate scenarios.

Our choices for the PRESTO project were the upper level ontology DOLCE (Descriptive Ontology for Linguistic and Cognitive Engineering) [7], and the classification of elements provided by XVR. DOLCE was chosen as this ontology not only provides one of the most known upper level ontologies in literature but it is also built with a strong cognitive bias, as it takes into account the ontological categories that underlie natural language and human common sense. This cognitive perspective was considered appropriate for the description of an artificial world that needs to be plausible from a human perspective. The decision to use the classification of elements provided by XVR was due to the extensive range of item available in their libraries (approximatively one thousand elements describing mainly human characters, vehicles, road related elements, and artifacts like parts of buildings) and the popularity of XVR as VR platform.

The construction of the first version of the ontology of PRESTO was therefore performed by following a middle-out approach, which combined the reuse and adaptation of the conceptual characterization of top-level entities provided by DOLCE and the description of extremely concrete entities provided by the XVR environment. More in detail,

- we performed an analysis and review of the conceptual entities contained in DOLCE-lite [7] together with the Virtual Reality experts (both trainers and developers) and selected the ones referring to concepts than needed to be described in a VR scenario; this analysis has originated the top part of the PRESTO ontology described in Sect. 4.1.
- we performed a similar analysis and review of the XVR items, together with their classifications, in order to select general concepts (e.g., vehicle, building, and so on) that refer to general VR scenarios; this analysis has originated the middle part of the PRESTO ontology described in Sect. 4.2.
- as a third step we have injected (mapped) the specific XVR items into the ontology, thus linking the domain independent, VR platform independent ontology to the specific libraries of a specific platform, as described in Sect. 4.3.

A reader could ask now why we didn't simply/mainly rely on the XVR classification in order to produce the, so called, PRESTO ontology. The reason is twofold: first of all, the XVR classification mainly concerns with objects. It provides therefore a good source of knowledge for entities "that are" (in DOLCE called Endurants), but a more limited source of knowledge on entities "that happen" (in DOLCE called Perdurants). Second, the XVR libraries contain objects described at an extremely detailed level whose encoding and classification resembles more to a Directory structures built to facilitate the selection of libraries rather than a well thought is-a hierarchy and therefore presents a number of problems that prevent its usage 'as such'. In the following, we review the most common problems we found in the categorization of the XVR items:

- Concepts names are used to encode different types of information. For instance the concept name "Caucasian_male_in_suit_34" is used to identify a person of Caucasian race, dressed in suit and of 34 years of age. Encoding the information on race, age, and so on via e.g., appropriate roles enables the definition of classes such as e.g., "Caucasian_person", "young adult", "male" and so on and the automatic classification (and retrieval) of XVR item via reasoning.
- The terminology used to describe concepts is not always informative enough: for instance, it is difficult to understand the meaning of the entity "HLO_assistant" from its label and description and to understand whether this item may suggest a type of "assistant" that may be useful in several scenarios and could therefore be worth adding to the ontology.
- The level of abstraction at which elements are described varies greatly. For instance the library containing police personnel items classifies, an the same hierarchical level the general concept of "Police_Officer" and the rather specific concept of "Sniper_green_camouflage".
- The criteria for the classification is not always clear: for instance, the "BTP_officer" (British Transport Police) concept is not a subclass of "Police_Officer".
- Certain general criteria of classification are not present in all the libraries. As an example, the general concept "Adult_Male" should be a general concept used for the classification of male characters. Nonetheless, it is present in e.g., the library of "Environment_humans" (that is, the library that describes generic characters) and is not present in e.g., the libraries of "Rescue_humans"

and "Victims" (that is, the libraries of characters impersonating rescuers and victims, respectively).

– Unclear classification: for instance, in the XVR original classification a "sign" is a "road_object", and a "danger_sign" is an "incident_object". By considering that no relations are defined between the entities "sign" and "danger_sign", it is not possible to infer any relation between "danger_sign" and "road_object".

– Duplication of concept names: for instance, the label "police_services" is used to describe both human police characters in the library "environment_human", and police vehicles, in the library "rescue_vehicle".

In the next section we provide an overview of the PRESTO ontology and of its top-level, middle level and XVR specific components in detail.

4   The PRESTO Ontology

As introduced in Sect. 3, the PRESTO ontology[3] is composed of three parts: (i) a top level part constructed with the help of DOLCE; (ii) a middle level describing general entities that can occur in a VR scenario, and (iii) a specific set of entities representing objects and "behaviors" available in a concrete VR.

4.1   The Top-Level Ontology: DOLCE Entities

Figure 1 shows the taxonomy of DOLCE entities taken from [7] revised and customised to the needs of PRESTO.

Entities in gray where not included in the PRESTO ontology, while entities in boldface where added specifically for PRESTO.

Among the first level of entities we selected **Endurants** and **Perdurants**: endurants are indeed useful to describe the big number of physical and non-physical objects that can occur in a serious game, including avatars, vehicles, tools, animals, roles and so on; perdurants are instead useful to describe what happens in a scenario. Concerning endurants the diagram in Fig. 1 shows the ones we selected to be included in PRESTO; note that we did not include the distinction between agentive and non-agentive physical objects because of an explicit requirement by the PRESTO developers. In fact, they require the possibility to treat every object in a VR as an agentive one for the sake of simplicity[4]. While perdurants can be useful in a VR to describe a broad set of "things that happen", in the current version of the ontology they were mainly used to describe animations (that is, "bodily_movements") of avatars. From an ontological point

---

[3] The current version of the PRESTO ontology cannot be published due to copyrights constraints. A preliminary version, from which it is possible to observe the rational used for modeling it, may be found here: https://shell-static.fbk.eu/resources/ontologies/CorePresto.owl.

[4] A typical example is vehicle, which the developers prefer to treat as an agentive objective, rather than a non agentive object driven by an agent, for the sake of simplicity of the code.

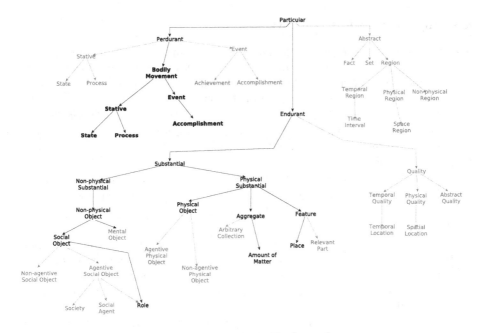

Fig. 1. The top-level PRESTO ontology.

of view we felt it was appropriate classify them according to the categories of stative and eventful perdurants included in DOLCE. In fact, we can have *state* bodily movements (e.g., being sitting), *process* bodily movements (e.g., running), and *accomplishment* bodily movement (e.g., open a door). The investigation of animations did not show examples of *achievement* bodily movements, which were therefore not included in the ontology.

The current version of the ontology does not contain **Qualities**, but current work (not described in this paper) is devoted to investigate how to include them in a further revision. Instead **Abstracts** do not seem to play a role in the PRESTO ontology.

4.2 The Middle-Level Domain Ontology

This part augments the top level ontology described above with concrete, but still abstract, entities that may appear in a broad range of VR scenarios for serious games. The current version of the ontology is composed of 311 concepts, 5 object properties and 3 annotations properties. Concerning the Endurant part the main entities modeled in the middle-level ontology pertain classifications of persons (avatars), buildings, locations, tools / devices, vehicles, and roles. Concerning perdurants the ontology contains concepts describing *state*, *process* and *accomplishment* bodily movement.

4.3   Injecting the Bottom-Level Ontology

The linking of the bottom-level ontology, representing the classification scheme used for organizing the items contained in the 3D-library, is not a trivial task. Indeed, the correct alignment of these levels enables the transparency of the system with respect to the actual content of the 3D-library.

While the creation of the top and middle-level of the PRESTO ontology is meant to create a stable knowledge source, the definition of the alignments with the bottom-level elements is an activity that has to be done every time a new 3D-library is plugged into the system.

To ease this injection we decided to accomplish it in two separate steps: (i) an automatic definition of alignments by using an ontology alignment tool and (ii) a manual refinement of the alignments before using the complete ontology in the production stage.

The output of the alignment task is the linking between the abstract concepts contained in the middle PRESTO ontology and the concrete items contained in the underlying 3D-library implemented in the system. Indeed, such alignments allow the access to the entire set of items defined in the 3D-library and that are physically used for building the VR scenario.

For sake of clarification about the alignment process works, let's consider the following example. In the middle-level of the ontology we have defined the concept "Tent" representing a general tent that may be used for building a VR scenario. By plugging, for example, the XVR library, we need to find an alignment between the entity "Tent" and the specific tent items contained in XVR, such as "Decontamination_Tent_Zone_1", "Family_tent_blue", "Treatment_Area", and so on. To do that, as first step, we execute the Alignment API library [8]: for the entity "Tent", the XVR item identified in the 3D-library and aligned with it is "Tents". Such an alignment, classifies the bottom-level ontology "Decontamination_Tent_Zone_1", "Decontamination_Tent_Zone_2", "Decontamination_Tent_Zone_3", "Family_tent_blue", "Family_tent_orange", "Festival_tent", and "Treatment_Area" as children of the concept "Tents". As a consequence, all these elements can be retrieved and used at run time to produce a specific scenario which requires the presence of a tent, while the scenario can still be described using the abstract term "tent". Also, the same high level scenario may be easily adapted to the usage of other 3D-libraries, simply by exploiting the (different) mappings of such libraries with the middle level "Tents" concept.

In some cases the automatic alignment we used fails: for example, the middle-level entity "Weapon" is automatically aligned with the bottom-level entity "Baton" instead of being aligned with the bottom-level entity "Service-weapon". In these cases, a manual refinement of the generated alignments was done afterwards for pruning wrong axioms.

By considering the XVR use case, the automatic alignment procedure allowed a time-effort reduction, with respect of doing everything manually, of around 65 % in the definition of the alignment between the middle-level and the bottom-level ontologies, thus showing the potential of using ontology mapping technologies in the concrete scenario of VR libraries.

5   Enriching the VR for Decision-Making and Coordination

There are a number of aspects required for decision-making and coordination of activities that cannot be fully captured via static taxonomies and aggregations but are worth describing in an ontology not only for its inherent representational and deductive power, which helps in structuring abstract reasoning, but for the ability built into PRESTO of dynamically and arbitrarily add and remove tags to any item within the VR. These tags are generically called "qualities" since they are mostly described as **Qualities** entities in the PRESTO ontology. They form a layer of knowledge shared by all PRESTO components (including configurator systems, DICE, an agent framework, agents, monitor and control GUIs, and end-user development tools) without the need of modifying the game engine or hard-coding relationships among categories and properties into software. Note that this layer could have been built into the ontology itself (technically, by representing all items in the VR as individuals stored in a triple store) but this would have created issues with distribution, deployment and performance, so it is managed differently. Further, DICE supports the tagging of BDI (Belief-Desire-Intention) plans and intentions by software developers; these tags can be used for introspection and monitoring of the activity of an agent.

Qualities are still work in progress, since they reflect the progressive development of behavioral models. At the moment, they are used for two main reasons: to represent an item's characteristics and dynamic state; and, to enable recognition (of activities and intentions) and coordination.

Examples of characteristics and states represented as qualities include:

- the characteristic of being a "gate", which indicates something that can be crossed but only after performing some enabling actions if required and coordinating with others, thus it is relevant to the models of navigation. A gate may be the revolving door at the entrance of a room, the sliding door of a lift, a driveway gate, a railroad crossing, and so on, all of which may have been classified very differently in the VR. Note that a permanently sealed door is not a gate in this definition;
- the dynamic state of being "open", which may be associated to gates (as above) as well as to entities not relevant to navigation (e.g. windows). Stative qualities are represented as a is-a hierarchy, whose root is a generic name (such as "openness") and whose children are the possible values of the quality (in this example, open, close, semi-open, semi-close, etc.). Items are tagged with the leaves (e.g., open or close) but the PRESTO API allows querying the current state by using the root, thus implicitly checking if the item does have that quality in the first place. Other examples of wide applicability include "liveliness" (which includes "alive", "dead", "impaired") and "functioning" (specialized in "running" and "stopped");
- dynamic states such as "body posture" and "facial expression", also organized in hierarchies as mentioned above. While posture and expression apparently are properties of humans only, they can be also applied to animals and even to

non-living entities; for instance, in shooting ranges (and their VR reconstructions), puppets used as targets may have different postures;
- dynamically changing values of various nature. PRESTO allows the association of an arbitrary content together with a tag to an item, thus this mechanism is essentially a way to add data fields to an object without impacting the general PRESTO API. For instance, the reward mechanism in a Unity game built for instructional purposes has been implemented as a "money"-tagged accumulator on a specific item.

As mentioned earlier, the PRESTO ontology classifies also the animations that can be applied by a game engine to entities. While this classification is used at the moment as a configuration tool, essentially to make DICE models agnostic with respect to the underlying technology, it is the first step towards a solution to the problem of intention recognition, which in turn is the base for the simulation of coordinated behaviour (no matter whether amicable, e.g. teamwork as fire fighters in the fire example presented earlier, hostile, e.g. opposition in a security scenario, or simply observation to anticipate future moves and take decisions, e.g. avoiding a safety exit door when too many people are engaging it during an alarm). Intention recognition is something that is innate in humans and cognitively complex animals (e.g. dogs) but computationally very hard if taken by principle; machine learning may come to the rescue in certain situations, but in a VR scenario where nuances of body and expressions are hard to capture and represent, let alone the limited number of training cases, this is not an option. In PRESTO, qualities are exploited to allow entities to make their recognizable activities publicly visible; thus, intention and action recognition is reduced to reading certain qualities automatically set by DICE when starting animations or appropriately tagged plans.

To do a further step ahead, work is in progress on game-theoretical descriptions of coordinated behavior, including queuing and other crowding behaviors, accessing shared resources, and so on, in order to enable the definition of policies at a very abstract (meta-) level. This work exploits, in addition to PRESTO's tagging of items, the equivalent in DICE for goals and plans as well as its support for introspection of intentions and motivations. In a nutshell, DICE agents tag themselves and any involved object with qualities that indicate the move they want to play in a coordination game, while their meta-level, cognitive models would try to achieve or stop pursuing aptly tagged goals and plans according to the agent's own moves in the game as well as of those entities perceived in the environment. The specification of policies is expected to substantially reduce the coding required by models and to allow the reuse of the same coordination patterns in many different situations, e.g. a single policy for queuing to pass through a gate (which will be part of the navigation models) as well as for queuing at the entrance of an office or at the cashier in a supermarket (which are decision-making behaviors not related to navigation goals).

A simplistic (but already available and of great practical use) coordinated behavior exploiting qualities is goal delegation from an agent to another agent. By means of the PRESTO API, any entity in a game can submit a goal to be pursued by any other entity; when the goal is enriched with a few predefined parameters, the destination DICE agent publishes the fact that it has accepted a goal or that has achieved it (or failed to achieve or refused), allowing the submitter (or any other observer, including PRESTO's session script engine) to monitor and coordinate behaviors without the use of any additional agent protocol.

In the PRESTO ontology, qualities are represented as endurant or perdurant, depending on their lifetime – static characteristics are endurant while stative, behavioral and coordination qualities are perdurant.

As a final note, it is worth mentioning that PRESTO uses ontologies, in addition to classifications and qualities as discussed above, for other purposes such as:

- to represent individual, rather than objective, perspectives on the world. Currently, an ontology is used to capture the possible values used by DICE models to appraise entities that may have an influence on behaviours. These values range from positive to negative at different levels, from "friendly" to "dangerous, to stay distant from". For reasons similar to those that led to the management of qualities in PRESTO, the relationships between ontological classifications and appraisal values are captured by configuration files at various level of granularity (shared by all NPCs of a certain type rather than specific for an individual) rather than within the ontology;
- software engineering practice, e.g. to allow the definition of certain APIs in a language-independent format, with the automatic generation of software in some cases, and similarly for independence from the game engine when accessing commonly available resource types (e.g. animations, as mentioned above) by means of an engine-neutral syntax.

6   Related Work and Conclusion

In this paper, we focused on the experience of using Semantic Web techniques, and in particular lightweight ontologies, for the description of the artificial entities and their behaviors in gaming with the aim of uncoupling the description of VR scenarios from their physical implementation in charge to the developers.

With respect to the literature, where ontologies are often used for a detailed description of the geometrical properties of space and objects [9], we focused more on how the description of the entities of a VR scenario can be easily represented and managed from the practical point of view. Indeed, the literature addressed such problems only marginally by focusing mainly on the use of ontologies for managing the representation of VR scenarios themselves [10,11], even if in some cases a clear target domain, like the management of information related to disasters [12], is took into account. Also the description of character behaviors have been supported by using ontologies for different purposes like as support for UML-based descriptions [13] or as a "core" set of structural behavioral concepts for describing BDI-MAS architectures [14].

However, all these works do not take into account issues concerning the practical implementations of flexible systems for building VR scenarios. The proposed solution demonstrated the viability of using Semantic Web technologies for abstracting the development of VR scenarios either from the point of view of the 3D-design and from the modeling of character behaviors.

References

1. Gruber, T.R.: Toward principles for the design of ontologies used for knowledge sharing? Int. J. Hum.-Comput. Stud. **43**(5–6), 907–928 (1995)
2. Guarino, N., Welty, C.: Identity and subsumption. In: Green, R., Bean, C., Myaeng, S. (eds.) The Semantics of Relationships: an Interdisciplinary Perspective. Kluwer, Dordrecht (2001)
3. Ritter, F.E., Bittner, J.L., Kase, S.E., Evertsz, R., Pedrotti, M., Busetta, P.: CoJACK: A high-level cognitive architecture with demonstrations of moderators, variability, and implications for situation awareness. Biologically Inspired Cogn. Architectures **1**, 2–13 (2012)
4. Evertsz, R., Pedrotti, M., Busetta, P., Acar, H., Ritter, F.: Populating VBS2 with realistic virtual actors. In: Conference on Behavior Representation in Modeling & Simulation (BRIMS), Sundance Resort, Utah, March 30–April 2 2009
5. Paternò, F.: End user development: survey of an emerging field for empowering people. ISRN Softw. Eng. **2013**, 1–11 (2013)
6. Fernández-López, M., Gómez-Pérez, A., Juristo, N.: Methontology: from ontological art towards ontological engineering. In: Proceedings of Symposium on Ontological Engineering of AAAI (1997)
7. Gangemi, A., Guarino, N., Masolo, C., Oltramari, A., Schneider, L.: Sweetening ontologies with DOLCE. In: Gómez-Pérez, A., Benjamins, V.R. (eds.) EKAW 2002. LNCS (LNAI), vol. 2473, pp. 166–181. Springer, Heidelberg (2002)
8. David, J., Euzenat, J., Scharffe, F., dos Santos, C.T.: The alignment API 4.0. Semant. Web **2**(1), 3–10 (2011)
9. Chu, Y.L., Li, T.Y.: Realizing semantic virtual environments with ontology and pluggable procedures. In: Lányia, C.S. (ed.) Applications of Virtual Reality. InTech, Rijeka (2012)
10. Bille, W., Pellens, B., Kleinermann, F., Troyer, O.D.: Intelligent modelling of virtual worlds using domain ontologies. In Delgado-Mata, C., Ibáñez, J. (eds.) Intelligent Virtual Environments and Virtual Agents, Proceedings of the IVEVA 2004 Workshop, ITESM Campus Ciudad de Mexico, Mexico City, D.F., Mexico, April 27th 2004, vol. 97 of CEUR Workshop Proceedings (2004). CEUR-WS.org
11. Xuesong, W., Mingquan, Z., Yachun, F.: Building vr learning environment: an ontology based approach. In: First International Workshop on Education Technology and Computer Science, ETCS 2009, vol. 3, pp. 160–165, March 2009
12. Babitski, G., Probst, F., Hoffmann, J., Oberle, D.: Ontology design for information integration in disaster management. In: Fischer, S., Maehle, E., Reischuk, R. (eds.) Informatik 2009: Im Focus das Leben, Beiträge der 39. LNI, vol. 154, pp. 3120–3134. Gesellschaft für Informatik, Bonn (2009)
13. Bock, C., Odell, J.: Ontological behavior modeling. J. Object Technol. **10**(3), 1–36 (2011)
14. Faulkner, S., Kolp, M.: Ontological basis for agent ADL. In: Eder, J., Welzer, T. (eds.) The 15th Conference on Advanced Information Systems Engineering (CAiSE 2003), Klagenfurt/Velden, Austria, 16–20, CAiSE Forum, Short Paper Proceedings, Information Systems for a Connected Society, CEUR Workshop Proceedings, vol. 74 (2003). CEUR-WS.org

Collaborative Editing of Ontologies Using Fluent Editor and Ontorion

A. Seganti[1(✉)], P. Kapłański[1,2], and P. Zarzycki[1]

[1] Cognitum, Wał Miedzeszyński 630, Warsaw, Poland
{a.seganti,p.kaplanski,p.zarzycki}@cognitum.eu
[2] Gdansk University of Technology, Narutowicza 11/12, Gdansk, Poland

Abstract. In this paper we present two tools that we are developing at Cognitum for managing large knowledge bases: Fluent Editor and the Ontorion Server. We have been able to build a collaborative knowledge management system using these two tools. We show how this system can be used for the concurrent modification of knowledge and how we can manage multiple modifications to the same knowledge.

1 Introduction

In this paper we show how to use Fluent Editor [1] and Ontorion [2,3] to build a collaborative editing tool for large ontologies that uses controlled natural language and modularization.

Fluent Editor is an editing tool for modifying ontologies using Ontorion Controlled Natural Language (OCNL) as an interface for editing. Fluent Editor's main features are: autocompletion - helping the user to write the correct sentences, many tools to interact with third party components (R plugins, Protégé plugin,...), an OCNL interface, reasoning and materialization, complex CNL queries to the current ontology, complete reference management (import/export/referencing of OWL/RDF ontologies), a graphical representation of the ontologies and interaction with the Ontorion Server. With its 2000+ users, Fluent Editor is quickly becoming an alternative to OWL-based editors like Protégé [4].

The Ontorion Server is the server equivalent of Fluent Editor, designed to have scalable reasoning, an OCNL interface (querying and saving), a SPARQL interface, OWL2/SWRL compatibility, tunable reasoning (currently OWL-DL and OWL-RL profiles are available) and high availability. Ontorion has advanced reference management, giving the user the possibility to define a prefix to a namespace map to be used for all entities. Inside Ontorion, an innovative modularization algorithm (based on [5]) is used to modularize the knowledge to allow for scalable reasoning.

Both Fluent Editor and the Ontorion Server expose an OCNL interface. OCNL is a controlled natural language designed on the one hand to be fully compatible with OWL2 and SWRL W3C standards and on the other hand, to be intuitive enough for people with little knowledge of logic to write knowledge bases. Internally, both products use description logic as the interface between these two worlds.

© Springer International Publishing Switzerland 2016
V. Tamma et al. (Eds.): OWLED 2015, LNCS 9557, pp. 45–55, 2016.
DOI: 10.1007/978-3-319-33245-1_5

In the first part of the paper we will briefly introduce Fluent Editor and the Ontorion Server. Then we will show the Collaborative Knowledge editing system architecture that we have built using these two tools and we will show how the modularization algorithm has been used to simplify knowledge modification. Next, we will present similar related collaborative ontology editors and compare them to our solution. Then, we will present the results of evaluation based on a pilot study of collaborative ontology editing in a Clinical Decision Support System application for Gist Cancer, followed by a discussion and a summary.

2   Fluent Editor and the Ontorion Server

Fluent Editor and the Ontorion Server are products created by Cognitum to manage knowledge. Both products are fully compatible with the W3C standards and internally use description logic for all logic-related operations. As the user interface for both systems, we use Ontorion Controlled Natural Language (OCNL), which is a controlled language equivalent to the OWL2 and SWRL languages.

2.1   Fluent Editor

Fluent Editor is a knowledge editor tool for editing standard W3C ontologies. In Fluent Editor, we provide: a reasoner, a taxonomy tree, a materialized graph, an interface to R programming using our ROntorion package, the possibility to import/export from/to the Protégé editor, a graphical representation of the ontology, reference management, a reasoning profile validation of the current ontology and Ontorion interoperability.

In Fluent Editor two different interfaces for ontology exploration have been implemented: a reasoner and a materialized graph. The reasoner uses an OWL-DL reasoner for reasoning over the content of the ontology. By default Fluent Editor loads HermiT [6] but it is possible to implement a simple C# interface to add other reasoners to Fluent Editor. In the materialized graph, we use a custom, Jena-based OWL-RL+ reasoner (OWL-RL+ is OWL-RL extended with some additional features based on the idea presented in [7] extended in a significant way with e.g. SWRL builtins, custom rules, etc.) to materialize the knowledge and then query the materialized inferred graph. In both cases, the user can make OCNL queries to the ontology and the results are displayed in OCNL. Due to differences in the underlying formalism, the query expressivity in the two cases is slightly different and the result window is also different: in the OWL-DL reasoner case, the result will consist of all entities answering the query together with sub and super concepts, while in the materialized graph case, the instances answering the query are shown.

2.2   Ontorion Server

The Ontorion Server [3] is a knowledge management server that offers scalable and tunable reasoning. Ontorion has an OCNL interface through which the user can make queries to the knowledge together with a SPARQL interface for complex SPARQL queries. It is possible to use the Ontorion Server through a Windows Communication Foundation (WCF) API, which exposes all the main functions to manage and query knowledge, manage users and databases and to use more advanced functionalities.

Reasoning in Ontorion is very similar to what is done in Fluent Editor's materialized graph. The main difference is that each time something is saved, Ontorion will extract the module of knowledge related to the sentences that are being saved and reason over this knowledge. This is possible because Ontorion has an implementation of a modularization algorithm (based on [5]) and allows reasoning to be scaled also for a large number of entities. In Ontorion two reasoning modes can be used: OWL-DL and OWL-RL+; however, in the first case, we still use the expensive OWL-DL reasoner, therefore the default mode for Ontorion is OWL-RL+.

In Ontorion, the administrator of the system can manage multiple knowledge bases and each user can have different access to the knowledge bases currently loaded. Furthermore, there is a knowledge versioning management feature to know when something has changed in the knowledge. This feature, together with optimistic concurrency over the implementation of modules, described here later, forms the basis for collaborative ontology editing.

2.3   OCNL

OCNL [8] is a controlled natural language for writing ontologies, which is compatible with OWL2+SWRL standards. Controlled natural language should be unambiguous and intuitive, ultimately forming an easy way for human-machine interaction (understandable by humans, executable by machines). Due to its limitations, it needs to be supported by a predictive (structural) editor and OCNL fulfills this requirements. It is currently implemented for English, Polish and German but can be extended to other languages. A typical sentence in OCNL will look like this: *Every man is a human-being.* In this sentence we can see that in OCNL complex words need to be separated by hyphens.

In OCNL it is possible to express all OWL constructs and SWRL rules (together with SWRL builtins) and use OWL references (prefixes *man[pfx]* or full namespaces *man*[<http://www.mynamespace.com>] can be used). General groups of sentences are allowed which include:

1. Concept subsumption, represents all cases where there is a need to specify (or constrain) the fact about a specific concept or instance (or expressions that evaluate the concept or instance) in the form of subsumption (e.g.: *Every cat is a mammal, Pawel has two legs* or *One cat that is brown has red eyes*).
2. Role (possibly complex) inclusion specifies the properties and relationships between roles in terms of the expressiveness of $\mathcal{SROIQ}^{(\mathcal{D})}$ (e.g.: *If X loves something that covers Y then X loves-cover-of Y*).
3. Complex rules; If [body] then [head] expressions that are restricted to the DL-Safe SWRL subset [9] of rules (e.g.: *If a scrum-master is-mapped-to a*

provider and the scrum-master has-streamlining-assessment-processes-sprints-level equal-to 2 then the provider has-service-delivery-level equal-to 1 and the provider has-support-services-level equal-to 2).

4. Complex OWL expressions; the grammar allows the use of parentheses that can be nested if needed in the form of (that <expression>) e.g.: *Every human is something (that is a man or a woman or a hermaphrodite).*

5. Knowledge modification triggers that have the form of: *If* P *then for-each* P *execute* Q, where P is a premise and Q a consequence. Premise P is an expression that evaluates a set of connected instances that fulfill some conditions, while the consequence Q is a procedure written in R [10] programming language.

2.4   System Architecture

The collaborative knowledge editing environment that we have implemented uses Fluent Editor and the Ontorion Server together. The architecture of the system is presented in Fig. 1.

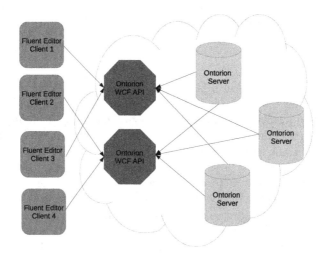

Fig. 1. Schema of the collaborative editing architecture using Fluent Editor and Ontorion.

The server side of the collaborative editing architecture is comprehensive combination of Ontorion and the Ontorion Windows Communication Foundation (WCF) API. The API exposes all Ontorion functionalities and has been used in Fluent Editor to implement the client-side. Ontorion has a complete user management system in which each user can have a different access level to each database in Ontorion. For collaborative editing, each user will connect using her/his Ontorion credentials, and the changes to the ontology will be logged.

On the client side, Fluent Editor is installed locally on the computer of each user. After the user opens Fluent Editor, s/he can connect to Ontorion and will enter into Fluent Editor in the Ontorion Mode. In this mode, it is possible to download/modify/add knowledge, see the taxonomy tree and make SPARQL queries to the knowledge in Ontorion.

2.5   Ontorion Mode

User interface. When in Ontorion mode Fig. 2, it is possible to Save, Download, Refresh and Clear the module regarding the knowledge the user is interested in. The whole point of the Ontorion mode is to manage modules of knowledge (see Sect. 2.6): the user adds entities to the signature, the module corresponding to the signature is downloaded and the user modifies or adds knowledge in the module.

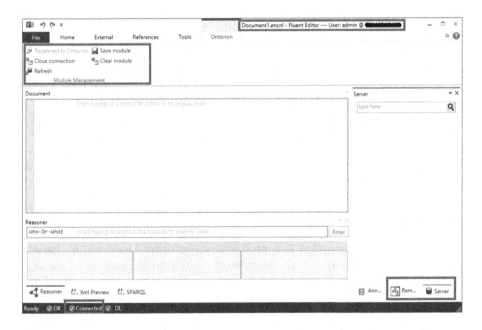

Fig. 2. Screenshot of the tab showing in Ontorion mode inside Fluent Editor.

Module management is based on the "signature" of the module. In the signature there are the entities (concepts, instances or properties) that are of interest to the user and each time a new entity is added, the module corresponding to the signature will be downloaded. For example, if the user adds to the signature the instance 'New-York' and the relation 'lives-in', the module will contain all knowledge relative to New-York and lives-in.

Taxonomy Tree and SPARQL queries. A taxonomy tree is displayed together with each ontology being edited in Fluent Editor and is generally built upon data from the current ontology and all referenced ontologies. Selecting an element on the taxonomy tree will search expressions in the CNL editor that are related explicitly to the selected element.

The taxonomy tree is divided into four parts:

- **thing**: shows *is-a* relations between concepts and instances.
- **nothing**: shows concepts that cannot have instances.
- **relation**: shows the hierarchy of relations between concepts and/or instances.
- **attribute**: shows the hierarchy of attributes.

In Ontorion mode, the taxonomy tree will load the structure of the knowledge currently stored in Ontorion. Furthermore, in Ontorion mode, it is possible to make SPARQL queries to the knowledge in Ontorion.

Annotation Support and References. Ontorion and Fluent Editor have full support for annotations and reference management. In Ontorion, it is possible to store a prefix to a namespace map that will be used to translate all namespaces contained in the entities to the corresponding prefix (e.g. label[rdf]). If no namespace is found, the entity will contain the full namespace (e.g. label[<http://www.w3.org/1999/02/22-rdf-syntax-ns>]). Thus each time that the user downloads the knowledge, Fluent Editor will also automatically download the references and the annotations related to the sentences that have been downloaded. It is then possible to see the annotations in the annotation window.

It is also possible to see annotations relative to all the elements in the taxonomy tree by right-clicking on one of the nodes and selecting the **Show annotation** command. In this case, the annotation will be shown without the need to download the knowledge relative to this entity.

2.6   Module Management

The modularization algorithm is a proprietary algorithm based on [5]. Essentially, a module in Ontorion is the knowledge relative to the entities in the signature together with the knowledge "around" it. This means that when asking for the module of an entity in Ontorion, the user will get back all the sentences where the entity is mentioned together with all the sentences that are related to this entity. Modularization is used internally by Ontorion to decide the part of the knowledge on which it needs to reason but, at the same time, it is used in the collaborative editing of knowledge to show the user only the knowledge she/he wants to work on.

2.7   Collaborative Knowledge Editing

The Ontorion mode is particularly interesting when used for collaboratively editing a big knowledge base stored in Ontorion. Indeed, more than one user can be connected to Ontorion at the same time and it is possible that both users are working on the same knowledge at the same time. In this case, when one of the users commits his/her changes to Ontorion, the other user will be notified of the changes as she/he will see the resync icon at the bottom left of Fluent Editor.

In this case, the user needs to click the Refresh button in the Ontorion tab, which will show her/him the sentences added from the other user to the module that is currently loaded (for example in Fig. 3 someone added a new attribute for the instance Eli). By clicking **Change**, this sentence will be added to the knowledge currently loaded into Fluent Editor.

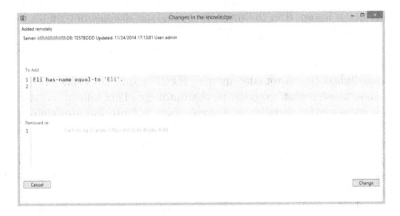

Fig. 3. Screenshot of the window that will be shown to the user when some changes are detected in the current knowledge that the user is editing.

When switched to "automatic merge" mode, if modules are frequently changing merging is done automatically without warning of non-colliding changes (separate lines of text w.r.t OCNL) turning the system into a real-time collaborative editing tool (RTCE).

In Fluent Editor (when in Ontorion mode), additional effort is made to manage module synchronization. In order to do this, we distinguish three modules: the current "local" module that the user is editing, the local remote module (the last module that the user downloaded) and the current remote module (the module that is currently stored in Ontorion). Using this distinction, we are able to identify the knowledge that has been added/removed locally to the module that was downloaded from Ontorion. Furthermore, in the background we check if the knowledge in Ontorion has changed. If it has we can, by comparing the remote module with the last downloaded module, understand if new knowledge has been added, and warn the user.

Moreover, a non-trivial problem related to managing module synchronization is caused by managing time synchronization. In order to avoid problems related to managing a common clock and other time-related problems in distributed systems, we opted for a simpler solution: knowledge versioning. Each time that a user makes a change to the knowledge base, the knowledge version changes. In this way, it is enough to download the knowledge version in Fluent Editor together with the module in order to know if the module we have downloaded is up-to-date or not.

This kind of conflict resolution can be described as "the first writer is the winner" synchronization. On the other hand, once the other user changes an overlapping knowledge module s/he is informed about it immediately, and is asked to update the state to the current version if a conflict occurs. This is very similar to such an approach common in software version control systems like e.g.: SVN; however, in SVN, conflict resolving is done during a so-called "workplace destabilization phase" and it is done on demand. Moreover, the programmer using SVN never knows if in the meantime someone else has not simultaneously changed the file. When working with Fluent Editor in Ontorion mode, the writer is always at least warned about a conflict; however, s/he can continue to work without an update at her/his risk.

When switched to "automatic merge" RTCE mode, all users are forced to work on the "most recent" version of the ontology. This kind of collaboration strategy is common for distributed systems (which Ontorion also claims to be) and can also be found e.g.: in Google Docs collaborative conflict resolution strategy, where only a critical conflict is reversed and the current merged state of the document is displayed, which can force the "late writer" to re-apply the change to the document [11]. From our observation, it seems to be true that this kind of conflict situation is very rare in common use cases.

This kind of conflict resolution greatly limits the need for communication during concurrent knowledge engineering and is possible due to the fact that users use controlled natural language as a medium. This is because the use of CNL makes the modern tools and methods used in concurrent document editing useful in the case of collaborative knowledge engineering.

Typically, collaborative editorial work requires additional channels of communication between editors that support the review process. As we have implemented full OWL annotation support in FE (also for OWL statements as annotation subjects), we assume that the whole review process can be implemented based on OWL annotations if only editors agree on a single annotation property to be such that describes "comments". Moreover, as all the changes are recorded, it is possible to track them if the review process requires it.

3   Related Work

CNL has a long history. A well-known implementation of a controlled natural language is "Attempto Controlled English (ACE)" [12], developed by the University of Zurich. However, the origins of CNL can be found in the famous novel by George Orwell: "1984", where he discusses NEWSPEAK – a controlled language. The most used industrial implementations nowadays are Domain Specific Language

(DSL) (implemented as a part of the Drools project) [13] and Semantics of Business Vocabulary and Rules (SBVR) [14]. Whereas CNL allows the representation of BPML diagrams. OCNL maps OWL2 in a bidirectional manner.

A wide range of collaborative ontology tools exists:

1. MoKi [15] is a tool that *"supports the creation of articulated enterprise models through structured wiki pages"*. It enables heterogeneous teams of experts, to collaborate. The active collaboration is guaranteed by an automatic translation between formal and informal specifications.
2. WebProtégé [16] is a free, open source collaborative ontology development environment. It is *"a simple editing interface, which provides access to commonly used OWL constructs"*. It supports full change tracking, revision history, sharing and permissions, threaded notes and discussions, watches and email notifications.
3. The NeOn [17] toolkit is *"a state-of-the-art, open source multi-platform ontology engineering environment, which provides comprehensive support for the ontology engineering life-cycle"*. It is based on the Eclipse platform and provides an extensive set of plug-ins covering a variety of ontology engineering activities, including a Collaboration Support plugin. This plugin allows one to manage the changes in different statuses, which are made by other editors.

In Fluent Editor, in Ontorion Mode, communication as well as ontology authoring is done based on CNL and OWL annotations, preferably in real-time when in "automatic merge", RTCE mode.

The use of a CNL for collaborative knowledge management can be found in AceWiki [18]. AceWiki is a semantic wiki that makes use of the controlled natural language ACE; the articles in the wiki are composed of formal statements that look like natural English.

While AceWiki is mostly focused on collaborative knowledge acquisition, we focus on the collaboration process between different experts and therefore on their intelligence amplification (IA); however, we are conscious of the blurred border between both approaches and we try to understand the properties of this border.

3.1 Community

Ontorion is free of charge for academic institutions and independent researchers. For more information, please visit our website available at http://www.cognitum.eu/semantics/.

4 Evaluation

The Clinical Decision Support System application for Gist Cancer (CDSS for Gist) was a pilot study devoted to Gastrointestinal Stromal Tumors (GIST), where we used collaborative ontology engineering based on our approach. Oncology is a field where recommendations are well defined and studied and where the quality of the clinical data needs to improve to allow for a more complex analysis of these data.

The experts involved were able to develop the ontology during this collaboration with the team of knowledge-engineering experts that had thought about the general architecture of the CDSS application and its non-functional requirements.

In this study we proved that if a domain expert is involved in the whole process of building the application then she/he is able to understand and correct the knowledge that has been written using OCNL.

5 Discussion and Summary

In this paper we presented a collaborative knowledge editing environment using Fluent Editor and the Ontorion Server. We have shown that using these tools it is possible to set up a collaborative knowledge editing environment where the user downloads the modules of knowledge that he/she is interested in and modifies the knowledge using OCNL language. Furthermore, the user can explore the knowledge currently stored in Ontorion by using SPARQL queries and the taxonomy tree.

We are currently working on improving the editing capabilities of the environment by: adding user access levels to each "module" of the knowledge (instead of the whole database), using knowledge modification logging to allow for undo functionalities, and adding versioning management.

References

1. Cognitum. Fluent Editor 2014 - Ontology Editor, 1 June 2015. http://www.cognitum.eu/semantics/FluentEditor/
2. Cognitum. Ontorion semantic knowledge management framework, 1 June 2015. http://www.cognitum.eu/semantics/ontorion/
3. Kapłański, P., Weichbroth, P.: Cognitum ontorion: knowledge representation and reasoning system. In: FEDCSIS 2015 (2015)
4. Knublauch, H., Horridge, M., Musen, M.A., Rector, A.L., Stevens, R., Drummond, N., Lord, P.W., Noy, N.F., Seidenberg, J., Wang, H.: The protege OWL experience. In: OWLED (2005)
5. Kapłański, P.: Syntactic modular decomposition of large ontologies with relational database. In: Nguyen, N.T., Katarzyniak, R.P., Janiak, A. (eds.) New Challenges in Computational Collective Intelligence. SCI, vol. 244, pp. 65–72. Springer, Heidelberg (2009)
6. Shearer, R., Motik, B., Horrocks, I.: Hermit: a highly-efficient owl reasoner. In: OWLED, vol. 432, p. 91 (2008)
7. Meditskos, G., Bassiliades, N.: Dlejena: a practical forward-chaining OWL 2 RL reasoner combining jena and pellet. Web Semant. Sci. Serv. Agents World Wide Web 8(1), 89–94 (2010)
8. Kapłański, P.: Controlled English interface for knowledge bases. Studia Informatica 32(2A), 485–494 (2011)

9. Glimm, B., Horridge, M., Parsia, B., Patel-Schneider, P.F.: A syntax for rules in OWL 2. In: Hoekstra, R., Patel-Schneider, P.F. (eds.) OWLED. CEUR, vol. 529. CEUR-WS.org (2008)
10. Ihaka, R., Gentleman, R.: R: a language for data analysis and graphics. J. Comput. Graph. Stat. 5(3), 299–314 (1996)
11. Dekeyser, S., Watson, R.: Extending google docs to collaborate on research papers. University of Southern Queensland, Australia 23, 2008 (2006)
12. Fuchs, N.E., Schwitter, R.: Attempto controlled English (ace) (1996). arXiv preprint cmp-lg/9603003
13. Proctor, M., Neale, M., McWhirter, B., Verlaenen, K., Tirelli, E., Bagerman, A., Frandsen, M., Meyer, F., De Smet, G., Rikkola, T., Williams, S., Truit, B.: Drools (2007)
14. OMG. Semantics of business vocabulary and business rules (sbvr), v1.0, Abruf am 02.05.2013 (2008). http://www.omg.org/spec/SBVR/1.0/PDF
15. Di Francescomarino, C., Ghidini, C., Rospocher, M.: Evaluating wiki collaborative features in ontology authoring. IEEE Trans. Knowl. Data Eng. 26(12), 2997–3011 (2014)
16. Tudorache, T., Nyulas, C., Noy, N.F., Musen, M.A.: Webprotégé: a collaborative ontology editor and knowledge acquisition tool for the web. Semant. Web 4(1), 89 (2013)
17. Haase, P., Lewen, H., Studer, R., Tran, D.T., Erdmann, M., d'Aquin, M., Motta, E.: The neon ontology engineering toolkit. In: WWW (2008)
18. Kuhn, T.: Acewiki: A natural and expressive semantic wiki (2008). arXiv preprint. arXiv:0807.4618

Integrating Ontology Negotiation and Agent Communication

Marlo Souza[1]([⊠]), Alvaro Moreira[1], Renata Vieira[2],
and John-Jules Ch. Meyer[3]

[1] Federal University of Rio Grande do Sul - UFRGS, Porto Alegre, Brazil
{marlo.souza,alvaro.moreira}@inf.ufrgs.br
[2] Pontifical Catholic University of Rio Grande do Sul - PUCRS,
Porto Alegre, Brazil
renata.vieira@pucrs.br
[3] Utrecht University, Utrecht, The Netherlands
j.j.c.meyer@uu.nl

Abstract. Ontologies are considered a necessary ingredient for communication among heterogeneous agents in the Web. With the multiplication of ontologies for the same domains, semantic interoperability has become a challenge. In this work, we study the use of ontology negotiation in a agent communication mechanism for agents with ontological reasoning. The resulting communication mechanism allows agents to exchange not only factual but also terminological knowledge about an individual domain and is closely related to available mechanisms in the literature such as KQML and FIPA-ACL.

Keywords: Ontologies for agents · Agent communication · Ontology negotiation

1 Introduction

It is commonly accepted that two essential ingredients for the construction of the Semantic Web are the use of ontologies and autonomous agents. Despite that fact, the integration of ontology-based reasoning in the semantics of communication mechanisms for multiagent systems has just recently became the focus of attention.

As different ontologies arise to describe the same domain, achieving *semantic interoperability* became essential to allow communication among agents in the Web. Various methods have been proposed to solve this problem in the area commonly known as Ontology Mediation [1].

The most popular approach to ontology mediation is ontology matching [2]. While it is a prolific area with mature methods, matching methods are static, i.e. the alignments are established before the agents' interactions. A dynamic alternative is to centralize ontology mediation in ontology agents [3]. The centralization required by this approach, however, is not easily scalable.

© Springer International Publishing Switzerland 2016
V. Tamma et al. (Eds.): OWLED 2015, LNCS 9557, pp. 56–68, 2016.
DOI: 10.1007/978-3-319-33245-1_6

Seeking dynamic and decentralized ways to ontology mediation, [4] proposed the notion of ontology negotiation. In this approach, the agents resolve their communication problems by negotiating a conversation vocabulary among themselves. We believe this approach is particularly interesting since it includes agent communication as an integral part of ontology mediation.

The main contribution of this work is a proposal to embed ontology negotiation into a speech act-based communication mechanism for multiagent systems. We introduce a mechanism to allow communication between agents with different private vocabularies that can exchange terminological information, based on the work of [5]. We will, however, drop the requirement of classifiers and thus generalize the conditions for meaningful communication between heterogeneous agents.

This work is structured as follows: in Sect. 2 we discuss the related work; in Sect. 3 we define the central ideas of the work of approximate translation and information loss and in Sect. 4 we establish formal conditions for proper communication between agents and algorithms for computing translations; in Sect. 5 we present a mechanism for ontology-based communication that allows ontology negotiation. We conclude the paper with some considerations about the problems that arise from integrating negotiation and communication in the language and the limitations of our technique.

2   Related Work

In [6] a framework for heterogeneous multiagent systems supporting ontology-based communication was defined. The framework allows agents to have different ontologies, and they implement an ontology service in which all the ontologies must be registered. The technique only allows assertional exchange and is centralized, by the use of ontology agents.

From the ontology negotiation point of view, [4] presents a protocol that allows agents to exchange parts of their terminologies and to interpret received messages. The communication mechanism, however, is very restricted. A similar approach but focused mainly on inductive methods used to locate similar concepts between the agents' terminologies is given in [7].

A method for semantic interoperability between taxonomies in peer-to-peer systems is presented in [8]. While their approach is very similar to ours, their work is limited to taxonomies, and they require a complete knowledge of the extension of the concepts.

The work of [5] presents different communication protocols for two agents to establish a communication vocabulary, by exchanging parts of their ontologies. They focus on how to preserve the extensional meaning of the concepts between different but jointly consistent ontologies. The main limitation of in [5] is that they assume that every agent knows the complete extension of their concepts - by the use of classification functions. Complete knowledge, however, is not a realistic property for most applications.

We generalized the notions of lossless communication of [5] for agents with incomplete information about their domain. Also, we integrate their terminological negotiation protocol within a communication mechanism for multiagent systems.

3   Approximated Translation and Information Loss

Our mechanism relies on two central notions: *approximate translation* and *information loss*. Informally, approximate translations are functions that transform a formula φ_s written using concepts of the speaker's ontology into a formula φ_h using concepts of the hearer's ontology in a way that preserves meaning. An information loss occurs when an agent detects that the meaning of an information may have been lost in the translation process.

From here on, we assume familiarity with Description Logics (DL). Capital letters such as A, B, C, C', etc. represent concept names while lowercase letters as a, b represent individuals. The uppercase Greek letter Φ_C will usually denote the set of concept names of an ontology, Similarly, Φ_R the set role names in an ontology. We will call $\mathscr{B}_{\Phi_i}$ the set of atomic concept in the ontology $\mathcal{O}_i$ - which is dependent on the DL chosen. For DL-Lite the foundational language of the OWL 2-QL profile, for example, it includes the concepts $\exists R$ for every role name $R \in \Phi_R$, besides the concept names.

We will not fixate a particular Description Logic, but only require that the negation of atomic concepts $\neg B$ to be representable in it. We are aware that some Description Logics do not allow atomic negation, such as the important EL family used as the foundation for the OWL2 - EL profile. Notice, however, that several important Description Logics, such as the DL-Lite family, SHOIN and SROIC, used to defined the semantic of parts of the Ontology Web Language, allow these constructions. We leave for future developments to extend our methods for DLs without atomic negation.

We require that the ontologies used by the agents are expressible in the same Description Logic. A more general setting would require the formal machinery of translations between logics [9], which is outside the scope of this work. We also require that their sets of concept names are disjoint. Concepts with the same name in different ontologies can be differentiated by the use of namespaces.

In our mechanism, each agent i has a concept translation function T_{ij}, for each agent j in the system. A concept translation function $T_{ij} : \mathscr{B}_{\Phi_i} \to \mathscr{B}_{\Phi_j}$ is a partial function that maps a concept $C_i \in \mathscr{B}_{\Phi_i}$ to a concept $C_j \in \mathscr{B}_{\Phi_j}$. A similar function may be defined for role names $Tr_{ij} : \Phi_{R_i} \to \Phi_{R_j}$. For sake of space we will not include translation of roles, but we point out that for expressible DLs, such as $SROIC$ [10] that bases the OWL 2 language, deciding translation for roles may be achieved using the same techniques as proposed below for concepts. We represent that the translation of C is undefined by $T_{ij}(C) = \bot$.

When a translation for a concept C in ontology $\mathcal{O}_i$ is undefined, the agent i will navigate in the hierarchy of concepts $\mathcal{O}_i$ searching for a concept $C' \neq C$ which is both "closer" to C in the hierarchy and is also translatable according to

T_{ij} (i.e. $T_{ij}(C') \neq \perp$). We call such a concept C' a most specific super concept of C, translatable according to a concept translation function T_{ij}.

Definition 1. *A most specific super concept of a concept $C \in \mathscr{O}_i$ translatable according to a concept translation function T_{ij} is an atomic concept $C' \in \mathscr{O}_i$ satisfying the following properties:*

(i) $C' \neq C$, $\mathscr{O}_i \models C \sqsubseteq C'$, $T_{ij}(C') \neq \perp$, and
(ii) for all $C'' \in \mathscr{O}_i$, if $C'' \neq C$, $\mathscr{O}_i \models C \sqsubseteq C''$, and $T_{ij}(C'') \neq \perp$ then $\mathscr{O}_i \not\models C'' \sqsubseteq C'$

We call $mstsc(\mathscr{O}_i, T_{ij}, C)$ the set of most specific super concepts of C translatable according to a concept translation function T_{ij}

By (i) in the definition above, C' is a superconcept of C, translatable w.r.t T_{ij}, while (ii) states that C' is most specific than any other concept C'' that is also a superconcept of C and translatable w.r.t T_{ij}.

Note that a most specific super concept always exists since axiom $C \sqsubseteq \top$ always holds, and we require that the concepts $\top$ and $\perp$ are always translated to themselves. Also, it might not be unique. An algorithm for computing the set of most specific super concepts of a concept C translatable according to a concept translation function T_{ij}, denoted by, is presented in Fig. 1.

Algorithm $mstsc(\mathscr{O}_i, T_{ij}, C)$
Input :
    ontology $\mathscr{O}_i$,
    translation function T_{ij} ,
    atomic concept C
Output :
    Set S of mstsc of the concept C
    [1] $S := \{C' \in \mathscr{B}_{\Phi_i} \mid C' \neq C, \mathscr{O}_i \models C \sqsubseteq C'$ and $T_{ij}(C') \neq \perp\}$
    [2] $S' := S$
    [3]   **for each** $C' \in S'$
    [4]   **for each** $C'' \in S'$
    [5]   **if** $\mathscr{O}_i \models C'' \sqsubseteq C'$ **then**
    [6] $S := S \setminus \{C'\}$
    [7]   **return** S

Fig. 1. Algorithm for computing the most specific translatable superconcepts of a concept

In the algorithm depicted in Fig. 1, line 1 computes all atomic concepts that satisfy condition i) in the Definition 1. In the loop (lines 3 to 6), the concepts that are not most specific, (i.e. that violates condition ii in Defintion 1) are discarded. We can now define the notion of approximated concept translation.

Definition 2. *Let $T_{ij} : \mathscr{B}_{\Phi_i} \rightarrow \mathscr{L}(\mathscr{B}_{\Phi_j})$ be a concept translation function, and $\mathscr{O}_i$ the agent's ontology. We define the approximated concept translation function $\overline{T_{ij}} : \mathscr{B}_{\Phi_i} \rightarrow \mathscr{B}_{\Phi_j}$ as:*

$$\overline{T_{ij}}(C) = \begin{cases} T_{ij}(C) & if \ T_{ij}(C) \neq \perp \\ T_{ij}(C') & otherwise, w/ C' = \mathsf{sel}(mstsc(\mathscr{O}_i, T_{ij}, C)) \end{cases}$$

The approximated translation of an atomic concept C is the concept translation of it, if it is defined, or it is the approximated concept translation of the most specific superconcept of C.

In the definition above, sel is a selection function on the set of most specific translatable superconcepts, which may be provided by the programmer.

Since most agent communication mechanisms in the literature [11–13] rely on communication of ground atomic facts, we will not deal with complex formulas. Our mechanism will be based only on instantiations of roles and concept literals, i.e. atomic concepts and their negations. Notice however that, given that the ontologies of the agents are defined over a same Description Logic, it is easy to lift the translation of atomic concepts to complex concept formulas by preserving the syntactic structure. The definition below may be, thus, generalized to be applicable in more expressive DLs.

Definition 3. *Let $C(a)$ be a concept instantiation, where C is concept literal formula in the vocabulary of the agent i. The translation of formula $C(a)$ to the terminology of agent j, represented by $\overline{T_{ij}}(C(a))$, is given by the following approximate formula translation function:*

$$\overline{T_{ij}}(C(a)) = \begin{cases} T_{ij}(C)(a)[Ex] & if\ T_{ij}(C) \neq \bot \\ \overline{T_{ij}}(C)(a) & otherwise \end{cases}$$

In the above definition, we use an annotated formula $T_{ij}(C)(a)[Ex]$ to point out explicitly that this is the product of an exact translation - this will be important later when we define information loss. From a logical point of view, these annotations have no interpretation.

When an agent i needs to communicate an information $C(a)$ to an agent j she will first translate $C(a)$ into j's terminology, using the function in Definition 3, and then send the message with the resulting formula.

We say that an agent i with ontology $\mathscr{O}_j$ and translation function T_{ij} satisfies $InfLoss(at)$, i.e. $\langle \mathscr{O}_j, T_{ji} \rangle \models InfLoss(at)$, when j detects a (possible) loss of information on receiving information at from agent j.

Definition 4. *We say the agent possessing ontology $\mathscr{O}_i$ and translation function T_{ij} detects an information loss when receiving an atomic formula at from agent j iff*

$$\langle \mathscr{O}_j, T_{ji} \rangle \models InfLoss(at)\ \ iff\ at \neq C(a)[Ex]\ and\ for\ some\ term\ t$$
$$\exists B \in \mathscr{B}_{\Phi_j} s.t. \overline{T_{ji}}(B(t)) = \overline{T_{ji}}(at)$$

When an agent detects an information loss, she can ask the speaker to clarify the information. This action will result in the speaker introducing new concepts in the communication, i.e. explaining to the hearer the meaning of the concept the speaker intend to use in the communication. Knowing the meaning of the concept the speaker wants to use, she can inform to the speaker an appropriate translation to her terminology.

4   Formal Properties of Translations Between Ontologies

The main problem for establishing a translation of a message is to decide the semantic relations between the concepts of the two ontologies. In this section, we introduce the formal properties required of the translation functions used in the mechanism described later. We will adopt the requirements of [5] of maximal preservation of extensional meaning as the guideline to decide translations. In this section, we aim to provide a function $TransCand$ to compute candidates for the translation of a concept literal C, as used in rule $AddConcept$ explained later.

As discussed earlier, we believe the limitations imposed in [5] are very restrictive, in the sense that they may not be applied to a wide range of MAS. Giving up these assumptions, however, implies we can no longer have certainty on which translation is the correct. We can identify only those which cannot be good translations since the computation may only be performed over an incomplete set of information about the world.

We assume a fixed set Δ of individuals for all ontologies, i.e. there are no private names for individuals. This assumption is equivalent to require individuals to be referenced by URIs. In the following, we define a series of properties that have to be satisfied by our translation between ontologies. These properties are based on those given in [5] posed in the context of incomplete information about the domain.

Definition 5 (Sound Translation). *Let $\mathscr{O}_i$ and $\mathscr{O}_j$ be two ontologies. We say $T_{ij} : \mathscr{B}_{\Phi_i} \to \mathscr{B}_{\Phi_j}$ is a sound translation from $\mathscr{O}_i$ to $\mathscr{O}_j$ iff*

$$\forall C \in \mathscr{B}_{\Phi_i}, \forall a \in \Delta \ (\mathscr{O}_i \vDash C(a) \Rightarrow \mathscr{O}_j \nvDash \neg T_{ij}(C)(a)).$$

Additionally, we say of any $C' \in \mathscr{B}_{\Phi_j}$ to be a sound translation for $C \in \mathscr{B}_{\Phi_i}$ if there is a sound translation T_{ij} from $\mathscr{O}_i$ to $\mathscr{O}_j$ such that $T_{ij}(C) = C'$.

The soundness condition means that the original meaning of the message is coherent with the translated message, i.e. the translation of an atomic concept encompass all the positive cases of the original concept. Since it cannot be guaranteed that there is no atomic concept in the target terminology with the same extension as the original concept, we consider a translation is sound if it is a superconcept of the original one, i.e. if it encompass all the positive information the original concept does.

Notice that we do not require, as [5], that the ontologies have complete information on all individuals. Consequently, by Definitions 5 (and 6 below), the computation of appropriate translations relies mainly on the shared individuals, i.e. the individuals that appear in both ontologies.

The other property required is that of lossless communication. To define that more elegantly, we will use the notion of extension of a DL formula φ in an ontology $\mathscr{O}$, meaning all the individuals that are inferred to be an instance of φ in $\mathscr{O}$, symbolically $ext(\mathscr{O}, \varphi) = \{a \in \Delta \mid \mathscr{O} \vDash \varphi(a)\}$.

Definition 6 (Lossless Translation). *Let $\mathscr{O}_i$ and $\mathscr{O}_j$ be two ontologies. We say $T_{ij} : \mathscr{B}_{\Phi_i} \to \mathscr{B}_{\Phi_j}$ is a lossless translation from $\mathscr{O}_i$ to $\mathscr{O}_j$ iff T_{ij} is a sound translation and for any atomic concept $C \in \mathscr{B}_{\Phi_i}$ there is no sound translation $T'_{ij} : \mathscr{B}_{\Phi_i} \to \mathscr{B}_{\Phi_j}$ such that*

$$ext(\mathscr{O}_i, C) \cap ext(\mathscr{O}_j, T_{ij}(C)) \subseteq ext(\mathscr{O}_i, C) \cap ext(\mathscr{O}_j, T'_{ij}(C))$$

and

$$ext(\mathscr{O}_i, \neg C) \cap ext(\mathscr{O}_j, T'_{ij}(C)) \subset ext(\mathscr{O}_i, \neg C) \cap ext(\mathscr{O}_j, T_{ij}(C)).$$

Similarly, we say of any $C' \in \mathscr{B}_{\Phi_j}$ to be a lossless translation for $C \in \mathscr{B}_{\Phi_i}$ if there is a lossless translation T_{ij} from $\mathscr{O}_i$ to $\mathscr{O}_j$ such that $T_{ij}(C) = C'$.

This property means that the translation of an atomic concept is the most specific translation possible in the target ontology. In other words, while the translation of an atomic concept may differ in extension from the original one since a complete preservation may not be possible, the difference in the extension is minimal (w.r.t. set inclusion).

The following properties state the meaning preservation properties of lossless translations. First we show that, up to extensional equivalence, the translated concept preserves the meaning of the original concept.

Proposition 1. *Let $\mathscr{O}_i$ be an ontology and T_{ii} a lossless translation from $\mathscr{O}_i$ to $\mathscr{O}_i$. Then the translation of any concept C is (extensionally) equivalent to C, i.e.*

$$\forall C \in \mathscr{B}_{\Phi_i} \ (ext(\mathscr{O}_i, C) = ext(\mathscr{O}_i, T_{ii}(C))))$$

It is easy to see that this proposition holds from the definition of lossless translation. Notice that C is maximal element concerning the properties of Definition 6. Since any other translation must include the extension of C, by maximality of C, their extensions must be equal.

The following proposition states that the information $T_{ij}(A) = B$, where T_{ij} is a lossless translation, can be identified as a (defeasible) subsumption axiom $A \sqsubseteq B$ in the union of the ontologies.

Proposition 2. *Let $\mathscr{O}_i$, $\mathscr{O}_j$ be ontologies, T_{ij} a lossless translation from $\mathscr{O}_i$ to $\mathscr{O}_j$ and $\mathscr{O}_T = \{C_1 \sqsubseteq C_2 \mid C_2 \neq \bot \wedge T_{ij}(C_1) = C_2\}$. Then $\mathscr{O}_i \cup \mathscr{O}_j \cup \mathscr{O}_T$ is consistent.*

Notice that, by definition of soundness, $O_1 \vDash C(a)$, then $O_2 \nvDash \neg T_{ij}(C)(a)$. Thus, there is no individual a s.t. $C(a)$ and $\neg T_{ij}(C)(a)$ are derivable from the ontology. For this reason, the proposition above holds.

From the definitions above, we can easily construct functions to compute the set of possible sound and lossless translations. We provide algorithms for the computation of those function (Fig. 2a and b), given that this is a central step in the strategy for ontology negotiation in our mechanism. Notice that other techniques in instance-based matching [2] can be easily integrated to order or select translations, as some similarity measure between concepts. Particularly,

Algorithm $CompSound(\mathcal{O}, Pos)$
Input :
   Knowledge base $\mathcal{O}$
   Set Pos of positive instantiations of a concept
Output :
   set of sound translations
   [1] $Sound := \{\}$
   [2] for each C concept name in $\mathcal{O}$
   [3]   if $Pos \cap ext(\mathcal{O}, \neg C) = \{\}$ then
   [4] $Sound := Sound \cup \{C\}$
   [5] return $Sound$

Algorithm $TransCand(\mathcal{O}, Pos, Neg)$
Input :
   Knowledge base $\mathcal{O}$
   Set of positive and negative instantiations Pos, Neg
   of a concept
Output : set S of lossless translations
   [1] $S := CompSound(\mathcal{O}, Pos)$
   [2] repeat
   [3] $S' := S$
   [4]   for each $C \in S'$
   [5]   for each $C' \in S' \setminus \{C\}$
   [6]   if $ext(\mathcal{O}, C) \cap Pos \subseteq ext(\mathcal{O}, C') \cap Pos$ and
   [7] $ext(\mathcal{O}, \neg C') \cap Neg \subset ext(\mathcal{O}, \neg C) \cap Neg$
   [8]   then $S := S \setminus \{C\}$
   [9] until $S = S'$
   [10] return S

(a) Algorithm for computing sound translations of a concept

(b) Algorithm for computing lossless translations of a concept

Fig. 2. Algorithms for computing translations

since a successful translation is dependent on shared individuals, other methods may provide alignments between individuals.

The algorithm $CompSound$ in Fig. 2a for computing admissible sound translations for an atomic concept C based on its positive and negative instantiations works by testing for each atomic concept in the hearer's ontology if the soundness condition is satisfied by this concept. If soundness is violated, the concept is rejected as a possible translation.

The algorithm $TransCand$ in Fig. 2b computes candidates for lossless translations of a concept C, based on its positive and negative instantiations. It works by, first selecting all atomic concepts in the knowledge base $\mathcal{O}$ that are sound translations for C and testing for each one if it satisfies the lossless condition of Definition 6. If this condition is violated, the concept is rejected as a possible translation. By the following result, we have that the algorithm presented in Fig. 2b is correct. Notice that since the set of concept and role names are finite, so it is the set of atomic concepts, and thus the algorithms presented always terminate.

Proposition 3. *Let $\mathcal{O}_i$ and $\mathcal{O}_j$ be two ontologies with $\mathcal{B}_{\Phi_i}, \mathcal{B}_{\Phi_j}$ their respective sets of atomic concepts, and $C \in \mathcal{B}_{\Phi_i}$ s.t. $ext(\mathcal{O}_i, \neg C) = Neg$ and $ext(\mathcal{O}_i \neg C) = Pos$. For all $C' \in TransCand(O_j, Pos, Neg)$, C' is a lossless translation of C in $\mathcal{O}_j$.*

It is not difficult to see that the algorithm is correct, since it test for every atomic concept of the ontology whether the requirements in Definitions 5 and 6 hold. Also, notice that the algorithm always terminates, since it is an iteration on a finite set of concepts.

5 Integrating Ontology Negotiation in Agent Communication

Once established the main notions used in this work, we begin the description of a communication mechanism allowing terminological negotiation. To specify such mechanism, we will use a simple model for an agent ag as a tuple consisted of is composed of an ontology $\mathcal{O}$, a collection translation functions T and a message base $M = \langle In, Out, Susp, Hist \rangle$ with the agent's messages Inbox, Outbox, Suspended Messages, and History of Messages, with all messages sent by the agent respectively. An agent is, thus, a triple $ag = \langle \mathcal{O}, M, T \rangle$.

When an agent $ag_i = \langle \mathcal{O}_i, M_i, T_i \rangle$, for example, wants to send a message to agent $ag_j = \langle \mathcal{O}_j, M_j, T_j \rangle$ she executes an action .send. Messages in the outbox M_{Out} have the form $\langle mid, id, ilf, cnt \rangle$, where mid is the message identifier, id is the hearer's identifier, ilf is the illocutionary force or type of the message and cnt its content. A message in the inbox M_{In} has the same format except that id is the identification of the agent that has sent the message.

A multiagent system is composed of agents $\langle ag_1, \ldots ag_n \rangle$ asynchronously communicating with each other. In a multiagent system with n agents, the component T of each agent i has $n - 1$ concept translation functions $T_{ij} : \mathcal{B}_{\Phi_i} \rightarrow \mathcal{B}_{\Phi_j}$, one for every other agent j.

The operational semantics of our communication mechanism is given by a set of rules that define a transition relation between configurations $\langle \mathcal{O}, M, T \rangle$. Intuitively the notation $\langle \mathcal{O}, M, T \rangle \longrightarrow \langle \mathcal{O}', M', T' \rangle$ means that, after one step in its execution, the components of agent $\langle \mathcal{O}, M, T \rangle$ may have been modified to $\langle \mathcal{O}', M', T' \rangle$

Since the main components of the configuration are tuples, we will use the subscript when referring to a specific component, e.g. M_{In} will be used to refer to the inbox In in the tuple M. We will also make use of *selection functions* that are defined by the agent programmer, e.g. the function S_M selects a message from the agent's message boxes, such as M_{In}, to be processed next.

We will assume that, unless negotiating the addition or explanation of a concept, the agents always translate their messages before sending them. This assumption can be easily implemented by taking the semantics of sending a message $\langle mid, id, ilf, cnt \rangle$ to automatically translate the contents of the message.

We begin the description of the performatives by explaining how the different ontologies affect the rules for assertional communication. Then, we introduce the main contributions of this work, i.e. the rules for terminological negotiation. We will explain the rules of the communication mechanism by instantiating them in the interaction between two agents (agent i and j).

Assertional Communication. Assertional communication refers to the communication about ground facts, such as proposed in [11,12]. Usual performatives available for the agents are ones as *Tell/Inform* for communicating to an agent some information and *Ask/Confirm* for querying an agent about certain information. In this work, we will limit our discussion to the *Tell* performative. The other performatives in the literature, such as those in [12], may be constructed similarly.

As in the ontology negotiation protocols described in [5] when the hearer detects a possible loss of information, she must proceed to request further explanation. In the simplest case when no loss of information occurs, the semantics of the communication is straight forward.

When an agent j receives an atomic formula $C_j(a)$ from agent i as a *Tell*, and there is no information loss, where C_j is a concept in agent j's terminology, then agent j must update her Abox with that information. We use an Update function to include a set new factual information in the Abox. While we do not provide a construction of such function, we point out some candidates have already been proposed in the literature, e.g. [14].

$$\frac{S_M(M_{In}) = \langle m_0, i, Tell, C_j(a) \rangle \quad \langle \mathcal{O}_j, T_{ji} \rangle \vDash \neg InfLoss(C_j(a))}{\langle \mathcal{O}_j, M_j, T_j \rangle \longrightarrow \langle \mathcal{O}'_j, M_j, T_j \rangle} \quad \text{(Tell)}$$

where:
$$\mathcal{O}'_j = Update(\mathcal{O}_j, C_j(a))$$

The more interesting case for us happens when an information loss has been detected. In these cases, the agent must request further clarification of the message. In our mechanism, we include the performative *ReqSpec* to represent this request. To request further specification of an information in the message, the agent will send a *ReqSpec* message with the information for which the agent has detected a possible information loss.

$$\frac{S_M(M_{In}) = \langle m_1, i, Tell, C_j(a) \rangle \quad \langle \mathcal{O}_j, T_{ji} \rangle \vDash InfLoss(C_j(a))}{\langle \mathcal{O}_j, M_j, T_j \rangle \longrightarrow \langle \mathcal{O}_j, M'_j, T_j \rangle} \quad \text{(TellInfLoss)}$$

where:
$$M'_{jOut} = M_{jOut} \cup \{\langle m_1, i, ReqSpec, C_j(a) \rangle\}$$

The request for a specification will begin an interaction for terminological exchange. While this terminological exchange is being performed, however, the assertional exchange that initiated it - the *Tell* message above - will be suspended, for it to be restarted after a new concept is introduced to express correctly the information agent i wished to convey.

Receiving a ReqSpec Message. The *ReqSpec* is aimed to request the expansion of the conversation vocabulary when a possible information loss is detected by the hearer. When agent i receives a *ReqSpec* message from agent j, with content $C_j(a)$, she must add a new concept to the vocabulary that will allow a lossless communication between the agents. This action is performed by sending *AddConcept* messages, explaining the (known) extensional meaning of the concept.

Notice that the information $C(a)$ that agent i initially wished to convey has been translated to $C_j(a)$ before the first message was sent. Because of that, to further explain it, agent i must reacquire the original information, stored in her message history.

Since a *ReqSpec* message initiates a negotiation process, the assertional message agent i wanted to send to agent j must be suspended, waiting the end of the negotiation. Agent i, thus, will remove it from her history and store it in her suspended messages, for it to be processed after the terminological exchange is over.

$$S_M(M_{In}) = \langle m_1, j, ReqSpec, C_j(a)[Tell]\rangle$$
$$\frac{\langle m_1, j, Tell, C(a)\rangle \in Hist}{\langle \mathcal{O}_i, M_i, T_i \rangle \longrightarrow \langle \mathcal{O}_i, M_i', T_i \rangle} \quad \text{(REQSPEC)}$$

where:
$$Hist' = Hist \setminus \{\langle m_1, j, Tell, C(a)\rangle\}$$
$$S = \{C(b) \mid \mathcal{O}_i \vDash C(b)\} \cup \{\neg C(b) \mid \mathcal{O}_i \vDash \neg C(b)\}$$
$$M_{iSusp}' = M_{iSusp} \cup \{\langle m_1, j, Tell, C(a)\rangle\}$$
$$M_{iOut}' = M_{iOut} \cup \{\langle m_1, j, AddConcept, S\rangle\}$$

Receiving an AddConcept Message. The *AddConcept* is aimed to inform the hearer of the (extensional) meaning of a new concept C to be used in communication. An agent may send an *AddConcept* as a means to introduce a new concept she wants to use or as a response to a request for a terminological specification.

When agent j receives from agent i an *AddConcept* message with the set S containing the extension to the concept C, she must search for the concepts in her ontology that constitute a good translation for this new concept and inform agent i this information. To compute the "good" candidates for the translation of C in terms of the hearer's concepts, we will use a function $TransCand$, presented in Sect. 4. In the rule, we use an auxiliary function S_T that selects one among the candidates for translation. This selection function may be provided by the programmer.

$$\frac{S_M(M_{In}) = \langle m_1, i, AddConcept, S]\rangle}{\langle \mathcal{O}_j, M_j, T_j \rangle \longrightarrow \langle \mathcal{O}_j, M_j', T_j \rangle} \quad \text{(ADDCONCEPT)}$$

where:
$$P = \{a \mid C(a) \in S\}$$
$$N = \{a \mid \neg C(a) \in S\}$$
$$B = S_T(TransCand(\mathcal{O}_j, P, N))$$
$$M_{jOut}' = M_{jOut} \cup \{\langle m_1, i, Translate, C \sqsubseteq B\rangle\}$$

Receiving a Translate Message. The *Translate* message is a response to a previous *AddConcept* message. It contains one information: a terminological axiom $A \sqsubseteq B$, where A is the concept she wants to use in communication and B is the translation computed by the sender of the *Translate* message to this concept. As a result, the receiver will update her translation function to include this new information.

$$\frac{S_M(M_{In}) = \langle m_1, j, Translate, A \sqsubseteq B]\rangle}{\langle \mathcal{O}_i, M_i, T_i \rangle \longrightarrow \langle \mathcal{O}_i, M_i', T_i' \rangle} \quad \text{(TRANSLATE)}$$

where:
$$T'(C) = \begin{cases} B & \text{, if } C = A \\ T(C) & \text{, otherwise} \end{cases}$$
$$M_{iSusp}' = M_{iSusp} \setminus \{\langle m_1, j, Tell, C(a)\rangle\}$$
$$M_{iOut}' = M_{iOut} \cup \{\langle m_1, j, Tell, T(C)(a)\rangle\}$$

6   Conclusions

In this work, we presented an integration of ontology negotiation and an agent communication mechanism. Using the notion of translation between ontologies, and the algorithms provided to compute such translations, we guarantee the communication to be meaning-preserving. It is important to notice that we focus on the integration of a negotiation protocol within a broader communication mechanism for agent communication not on a method for ontology mediation *per si*.

About the translation method, it is important to notice two things. Firstly, by giving up on the use of classifiers as in [5], the successful communication between agents relies on the existence of shared individuals in the agents' ontologies. Secondly, it is also important to notice that the choice to translate concept names into concept names has expressibility consequences. Allowing the result of a translation to be a DL formula would increase expressibility of the method, as shown in [8]. The choice we made in our work was based on the fact that most communication mechanisms available rely on the exchange of ground literals, not complex formulas and that allowing complex formulas as translations - even if restricted to conjunctions of atomic concepts - yields in an exponential complexity for computing the translation candidates. Also, we would like to point out that a method of translation from concept names to formulas is highly dependent on which DL is used to axiomatize the ontologies while our method is general.

Regarding complexity, our approach requires only a linear number of query answering requests for the ontology reasoner. If the underlying Description Logic is limited enough, the computation of translations is tractable. We don't consider the integration of ontology negotiation in the communication mechanism to introduce a considerable overhead to the system. The reason for this belief is that translations are cumulative throughout the execution and are only computed when needed. This leads us to conclude that our method is scalable to large and open-ended systems, without creating a great overhead.

In future work, we would like to explore more deeply the connection between translation functions and defeasible subsumption rules. We believe the semantics developed for defeasible description logics may provide a rich understanding of ontology negotiation as a reasoning problem.

References

1. de Bruijn, J., Ehrig, M., Feier, C., Martín-Recuerda, F., Scharffe, F., Weiten, M.: Ontology mediation, merging and aligning. In: Semantic Web Technologies, pp. 95–113 (2006)
2. Euzenat, J., Shvaiko, P.: Ontology Matching, vol. 18. Springer, Heidelberg (2007)
3. FIPA Ontology service specification: FIPA XC00086D (2001)
4. Bailin, S.C., Truszkowski, W.: Ontology negotiation between intelligent information agents. Knowl. Eng. Rev. **17**(1), 7–19 (2002)
5. Van Diggelen, J., Beun, R.-J., Dignum, F., Van Eijk, R.M., Meyer, J.-J.: Ontology negotiation: goals, requirements and implementation. Int. J. Agent-Oriented Softw. Eng. **1**(1), 63–90 (2007)

6. Mascardi, V., Ancona, D., Bordini, R.H., Ricci, A.: CooL-agentspeak: enhancing agentspeak-DL agents with plan exchange and ontology services. In: Intelligent Agent Technology - IAT, pp. 109–116 (2011)
7. Williams, A.B.: Learning to share meaning in a multi-agent system. Auton. Agents Multi Agent Syst. **8**(2), 165–193 (2004)
8. Tzitzikas, Y., Meghini, C.: Ostensive automatic schema mapping for taxonomy-based peer-to-peer systems. In: Klusch, M., Omicini, A., Ossowski, S., Laamanen, H. (eds.) CIA 2003. LNCS (LNAI), vol. 2782, pp. 78–92. Springer, Heidelberg (2003)
9. Carnielli, W.A., Coniglio, M.E., D'Ottaviano, I.M.: New dimensions on translations between logics. Logica Universalis **3**(1), 1–18 (2009)
10. Horrocks, I., Kutz, O., Sattler, U.: The even more irresistible SROIQ. In: KR, vol. 6, pp. 57–67 (2006)
11. Finin, T., Fritzson, R., McKay, D., McEntire, R.: KQML as an agent communication language. In: International Conference on Information and Knowledge Management, pp. 456–463. ACM (1994)
12. Vieira, R., Moreira, A., Wooldridge, M., Bordini, R.H.: On the formal semantics of speech-act based communication in an agent-oriented programming language. J. Artif. Intell. Res. **29**(1), 221–267 (2007)
13. Klapiscak, T., Bordini, R.H.: JASDL: a practical programming approach combining agent and semantic web technologies. In: Baldoni, M., Son, T.C., van Riemsdijk, M.B., Winikoff, M. (eds.) DALT 2008. LNCS (LNAI), vol. 5397, pp. 91–110. Springer, Heidelberg (2009)
14. Liu, H., Lutz, C., Milicic, M., Wolter, F.: Updating description logic aboxes. In: KR, pp. 46–56 (2006)

Lifting EMMeT to OWL Getting the Most from SKOS

Bijan Parsia[1], Tahani Alsubait[1], Jared Leo[1(✉)], Veronique Malaisé[2],
Sophie Forge[2], Michelle Gregory[2], and Andrew Allen[2]

[1] The University of Manchester, Manchester, UK
{bijan.parsia,tahani.alsubait,jared.leo}@manchester.ac.uk
[2] Elsevier B.V., Philadelphia, USA
{v.malaise,s.forge,m.gregory,a.allen}@elsevier.com

Abstract. SKOS and OWL are quite different but complimentary languages. SKOS is targeted at "cognitive" or "navigational" representations, that is, thesauri, controlled vocabularies, and the like. OWL is targeted at logical representations of conceptual knowledge. To a first approximation, SKOS vocabularies try to capture *useful* relations between concepts, whereas OWL ontologies aim to capture *true* relations between concepts. Now, of course, the true is sometimes useful and the useful often true, thus SKOS and OWL overlap to some degree. However, there are applications where we need to know true relations (e.g., generating multiple choice questions). Furthermore, SKOS relations are not precisely specified (by design). For example, many different ways of being useful can be covered by the same SKOS relation, but only one way of being useful is actually applicable to some application.

In this paper, we present a case study of modifying a large, existing SKOS vocabulary partially into OWL. This lifting is motivated by an application (generating multiple choice questions) that requires more precision in the representation than SKOS alone supports.

1 Introduction

A central use case for Web Ontology Language (OWL) Ontologies has been the development and maintenance of "terminologies" such as controlled vocabularies, taxonomies, or thesauri. Indeed, many of the most significant (in terms of longevity, funding, use and size) ontologies such as SNOMED CT [15,16], the NCI Thesaurus,[1] or the Gene Ontology [3] are exemplars of this use case. The use of *reasoners* to support development time services such as debugging and verification [14] as well as runtime services such as post-coordination [5] is a key part of the OWL success story for these use cases. For these, the existence of a precise formal semantics for such constructs as SubClassOf is a boon.

However, not all relations of interest for terminologies fit into the OWL model. Indeed, many controlled vocabularies do not need any such precision.

---

[1] http://ncicb.nci.nih.gov/NCICB/core/EVS.

V. Tamma et al. (Eds.): OWLED 2015, LNCS 9557, pp. 69–80, 2016.
DOI: 10.1007/978-3-319-33245-1_7

Thus, OWL is sometimes both *too strong* (e.g., A `SubClassOf` B makes an onto-logical commitment — that every instance of A is an instance of B — which may be the wrong one for the hierarchical relationship we want) and *too weak* (e.g., we do not have ways to indicate that terms are related *in some way or another*, at least not very easily). Hence, the introduction of the Simple Knowl-edge Organisation System – SKOS [10].

SKOS takes the opposite approach than OWL. Instead of insisting on a formal semantics, SKOS has an informal (or perhaps semi-formal) semantics,[2] at least for domain knowledge. SKOS is designed for representing loose *cognitive* or *navigational* relations rather than accurate domain relations. Roughly, OWL aims to model *the way the world* is whereas SKOS aims to support *how we think about it* in some context. It is easy to see how the two might come apart. Consider the mnemonic for spelling 'ocean': "Only Cats' Eyes Are Narrow". In a knowledge organisation system (KOS) about *mnemonics*, we strongly associate 'ocean' and 'cats' even though there are few domain relations between them.

Thus, SKOS allows for some flexibility and ambiguity that is a good match for many applications (such as query expansion) where the application's seman-tic demands are similarly loose. However, SKOS representations can become implicitly overfitted to some applications as they co-evolve. Furthermore, there are some applications where ontological representations are better suited. But migrating from a SKOS representation to an OWL one is challenging at best. As the SKOS Reference states [10]:

> To make the "knowledge" embedded in a thesaurus or classification scheme explicit in any formal sense requires that the thesaurus or classification scheme be re-engineered as a formal ontology. In other words, some per-son has to do the work of transforming the structure and intellectual con-tent of a thesaurus or classification scheme into a set of formal axioms and facts. This work of transformation is both intellectually demanding and time consuming, and therefore costly. In addition, some KOS are, by design, not intended to represent a logical view of their domain. Con-verting such KOS to a formal logic-based representation may, in practice, involve changes which result in a representation that no longer meets the originally intended purpose.

---

[2] Take, for example, this paragraph from the SKOS Reference: "To understand this distinction, consider that the "knowledge" made explicit in a formal ontology is expressed as sets of axioms and facts. A thesaurus or classification scheme is of a completely different nature, and does not assert any axioms or facts. Rather, a thesaurus or classification scheme identifies and describes, through natural language and other informal means, a set of distinct ideas or meanings, which are sometimes conveniently referred to as "concepts"... These structures, however, do not have any formal semantics, and cannot be reliably interpreted as either formal axioms or facts about the world. Indeed they were never intended to be so, for they serve only to provide a convenient and intuitive map of some subject domain, which can then be used as an aid to organising and finding objects, such as documents, which are relevant to that domain."

We present a case study of ontologising a large clinical SKOS terminology, the Elsevier Merged Medical Terminology (EMMeT). EMMeT was initially released as a SKOS knowledge base. The main rationale was publish the vocabulary in a standard format for publication on the Web. Given EMMeT's non-formal structure, SKOS was a more fitting choice than OWL, in terms of standards, and was fitting for the first use cases of browsing and query expansion.

While EMMeT (as a KOS) is an excellent resource for current applications, it is not by itself suited for our application, to wit, the generation of multiple choice questions (MCQs). In particular, our MCQ generation technique requires us to distinguish between *true* and *false* subclass relations. In this paper, we describe our attempts to partially re-engineer EMMeT into an OWL Ontology.

2 Preliminaries

A *controlled vocabulary* is a collection of terms, possibly with their informal definitions. A *classification* is a controlled vocabulary that is usually, but not necessarily, hierarchically ordered. It provides a similarity-based grouping of concepts with respect to certain agreed principles, e.g., which similarity notions will be used for classifying the concepts. A *thesaurus* is a collection of concepts that can be related in three main kinds of relations: broader_than, narrower_than and related_to. The first two relations can be used to provide an informal hierarchical order of concepts while the third relation can be used to capture some notion of relevance for a given purpose, e.g., `Cars` are related_to `Fuel`. A thesaurus can also have synonymy relationships to allow for terminological level modelling. A *taxonomy* is also a collection of concepts but it is different from a thesaurus in terms of the underlying relations. In particular, a taxonomy is built using the so-called is_a relation which can provide a real subsumption hierarchy.

SKOS [10] is a World Wide Web Consortium (W3C) recommendation since 2009. It provides a lightweight language for representing knowledge in controlled structured vocabularies, classifications or thesauri and can be encoded using any concrete RDF syntax, e.g., RDF/XML. SKOS concepts can be linked to other SKOS concepts using hierarchical (e.g., `skos:narrower` and `skos:broader`) or associative relations (e.g., `skos:related`). In SKOS, the relation `rdf:type` which is used to specify instances of concepts is not available. Thus, extensional connections between concepts are not modelled. For example, consider a SKOS concept about "computers" which can be associated with the concept "printer" via a broader_than relation. Clearly, not all instances of printers are also instances of computers; hence, this is not a valid subsumption relation. However, the concept "computers", in this context, may be interpreted as computers and related devices from a sales and marketing perspective. As a result, SKOS hierarchical relations are not transitive by default. For example, a "printer" might be related to, indeed broader than, "A4 paper" but, "A4 papers" are not necessarily narrower than "computers" (via "printer" being narrower than "computers") because while "A4 paper" might well be a natural more specific search from "printer", it might not be a reasonable next level search from "computers": people looking at printers often want to buy paper. Fewer want to shop for paper while considering computers. Navigationally, there is a chain but we don't want shortcuts through that chain.

The Web Ontology Language (OWL) is the W3C standard ontology language for the web and was standardised in 2004. It provides a formal knowledge representation language with unambiguous semantics, i.e., context-independent meaning. An OWL ontology is a finite set of axioms that describe the main notions, i.e., concepts, of a domain of interest. The inferred class hierarchy is the Hasse diagram of the partial order on concept names, e.g., A and B, in an ontology $\mathcal{O}$ induced by the entailment relation $\mathcal{O} \models A \sqsubseteq B$. The main relation in inferred class hierarchies is the is_a relation which is a transitive subsumption relation. Clearly, this is different from skos:narrower and skos:broader relations which are not necessarily valid subsumption relations. OWL exploits Description Logics (DLs) [1] to provide ontologies with formal semantics.

Part of OWL's success is the availability of a number of optimised reasoners such as FacT++ [17], Pellet [14], HermiT [13], and ELK [7].[3] Different ontology editing and processing tools and libraries are readily available as well such as *Protégé*,[4] and the OWL API [2].

In addition to the standard reasoning services provided by the above reasoners, some useful non-standard reasoning services have also been developed. For example, many techniques have been developed to extract modules, i.e., subsets of the axioms in a given ontology that are "relevant" to a particular signature. An interesting property of modules is that they preserve all entailments relevant to the intended signature, yet they are much smaller than the original ontology. For example, if $\mathcal{O} \models C \sqsubseteq E$, where E is a concept expressible in a considered DL, then extracting a module $\mathcal{M}$ with a seed signature $\Sigma = \{C\}$ also guarantees that $\mathcal{M} \models C \sqsubseteq E$.

The Elsevier Merged Medical Taxonomy (EMMeT) is currently modelled in SKOS. It contains 927,827 concepts with 3,010,262 synonyms in the EMMeT 3.8 release (May 15 2015). Some prominent such areas include: Anatomy (17,000 concepts), clinical findings (8,500), drugs (40,500), organisms (34,000), procedures (61,000), along with symptoms (38,000). Further more there are 132,000 semantic relationships between these concepts.

EMMeT uses 3 types of elements in their SKOS representation: skos and skosxl elements, custom nodes used to represent semantic relations, and meta data nodes. Amongst the skos and skosxl terms are elements to classify concepts, e.g., skos:Concept, skos:ConceptScheme, skosxl:prefLabels, and elements used to act as relations between concepts such as skos:narrower and skos:broader and skos:ExactMatch which expresses a relationship between concepts from EMMeT and external conceptSchemes or vocabularies. Whenever possible, EMMeT makes use of existing standard properties, like the Dublin Core set for metadata, the PROV vocabulary for Provenance, RDF, SKOS and SKOS-XL. Whenever a custom property is needed, like explicit semantic relationships (which are more precise than skos:related), the idea is to create

---

[3] For a list of DL reasoners: http://owl.cs.manchester.ac.uk/tools/list-of-reasoners/.
[4] http://protege.stanford.edu.

them as sub-properties of standard W3C properties, to keep the compatibility with other published vocabularies. The metamodel, however, was not published together with the EMMeT release as it should have been.

The namespace `semrel` (semantic relation) was used in order to represent a concept to concept relation and specify a *ranking* of importance that the concept to concept relation has in the general knowledge base. For example,

```
<semrel:isACauseFor rdf:ID="Relation-2996187-i".../>
```

defines a relationship between two concepts, with an ID that allows for the reification of this relationship. The reification method is used to assign a rank to that relationship. For example:

```
<semrel:Relation rdf:about="...">
    <semrel:rank>9.0</semrel:rank>
</semrel:Relation>
```

These ranks are used in several ways including to filter or order results. For example, a very low ranked related concept might only be displayed if no more high ranked related concepts are found.

`emloc` (EMMeT local) is another defined namespace which represents a very specific semantic relationship between a coordinated concept (e.g., disease due to X symptoms) and its compounds: "Disease due to X" and "Symptoms". This relationship was designed to be used in a very specific case of knowledge intensive query expansion: to link a disease to its symptoms, treatments etc. and allow for the actual symptom/treatment concepts to be added to the query expansion.

```
<emloc:hasLocalChildren rdf:ID="Relation-3041760-h".../>
<!--  Disease due to Deltaretrovirus Symptoms  -->
```

There are also other metadata properties that are used as provenance and quality assurance for concepts creation and maintenance, for example information representing creation dates (`pav:createdOn`).

To illustrate the use of these nodes, consider Fig. 1. This example shows an extraction of EMMeT highlighting the usage of the elements described above. The example represents a graph between the 5 following concepts:

- `breastCancer`
- `malignantMelanomaOfBreast`
- `malignantMelanomaOfSkinOfBreast`
- `oncology`
- `radiationTherapy`

4 narrower/broader relations:

- `<breastCancer>skos:broader<malignantMelanomaOfBreast>`
- `<malignantMelanomaOfBreast>skos:narrower<breastCancer>`
- `<malignantMelanomaOfBreast>`
  `skos:broader<malignantMelanomaOfSkinOfBreast>`

– `<malignantMelanomaOfSkinOfBreast>`
  `skos:narrower<malignantMelanomaOfBreast>`

and 2 ranked semantic relations:

– (`<malignantMelanomaOfBreast>`
  `semrel:hasPhysicianSpeciality<oncology>`)`rank:6.0`
– (`<malignantMelanomaOfBreast>`
  `semrel:hasTreatmentProcedure<radiationTherapy>`)`rank:6.0`

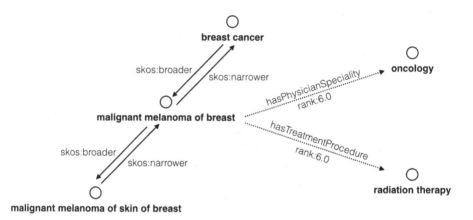

Fig. 1. An extraction from EMMeT, illustrating the use of concepts and their relations

3   Lifting EMMeT to OWL

EMMeT is clearly very large and has coverage on the scale of other major clinical terminologies. It is not feasible to do a hand crafted lifting based on term by term view by clinically savvy medical experts. We needed at least a "good enough" initial translation that hopefully will be sufficient for assessing whether EMMeT is in principle adequate for our task. To this end, we first attempted what seemed to be a straightforward, if a bit naive approach, on the theory that EMMeT might be close enough to OWL, and if not, such a translation would flush out problems. The latter proved to be the case, so we performed a more sophisticated but still automatic translation making use of the fact that many EMMeT concepts are mapped to SNOMED.

3.1   The Naive Approach

The basic idea of the naive approach is to presume that EMMeT's use of SKOS is more or less as a kind of syntax for OWL.

Classes: Clearly, on nearly any translation, we will map `skos:Concept` to `owl:Class`. In general, `owl:Classes` are intended to represent concepts and act as a way to describe part of a domain. Since `skos:Concepts` are used to describe a particular part of a knowledge base, or a term which covers other terms of similar properties, the mapping to `owl:Class` seemed obvious.

Relations: EMMeTs `semrel:X` relations are associative relations intended to relate concepts via a property. Using the example from Sect. 2, (`malignant MelanomaOfBreast semrel:hasTreatmentProcedure radiation Therapy`) `rank:6.0` is intended to represent that a *malignant melanoma of the breast* has a *treatment procedure* that is *radiation therapy*. `skos`'s `semantic relations` are described in the documentation as equivalent to `owls object propertys`. Since both `radiationTherapy` and `malignantMelanomaOf Breast` are classes, this leads to the following OWL modelling choice:

`malignantMelanomaOfBreast` $\sqsubseteq$
$\quad\quad$ `∃hasTreatmentProcedure.radiationTherapy`

where `hasTreatmentProcedure` is an instance of `owl:objectProperty`. In OWL, at an axiom level there are two ways to model property relations between two concepts. The first being universal ($\forall$) restrictions, and the second, as used in the example, being existential ($\exists$) restrictions. To illustrate the basic use of each example, consider the following two axioms

$$A \sqsubseteq \exists R.D \quad (1)$$
$$A \sqsubseteq \forall R.D \quad (2)$$

(1) says that every instance of A has an R-successor to an instance of D, whilst the second tells us that every instance of A must only have R successors to instances of Ds. Since we don't want to enforce the restriction that `malignantMelanomaOfBreast` only has `radiationTherapy` as a treatment procedure and no other treatment (EMMeT also tells us that `simple mastectomy`, `chemotherapy` and `mastectomy` are also treatment procedures with ranks `5.0`, `6.0`, `6.0` respectively), we opted for the existential axiom, leaving the door open for other treatments being allowed. The same modelling options were used for EMMeT's relations `emloc:hasLocalChildren` and its inverse `isLocalChildOf`. Although these may be interpreted as similar to narrower and broader relations, after closer inspection this was not so the case, therefore `owl:objectPropertys` were used, in the same way as above. In general, existential readings are preferred in biomedical ontologies not least because it opens the possibly of staying inside the polynomial OWL EL profile, which is the profile of choice for SNOMED, among others.

Narrower and Broader: `skos:narrower` and `skos:broader` are hierarchical relations used to indicate that one concept is in some way more general (*broader*) than another concept (*narrower*). We chose to model these using OWL's `subClassOf` relation. There is really no other reasonable choice in OWL if we want to model hierarchical concept relations, at least, without a complex scheme that is unlikely to be understood by other modellers. Of course, we know that `subClassOf` is transitive whereas the source SKOS relations are not. However, it might be that they *are* transitive when read ontologically. After all, it is a standard to use the transitive reduct of an OWL class hierarchy as a navigational structure — this is exactly what the Protégé class hierarchy view does.

Ranks. In EMMeT, `semrel:rank` properties were used to map associative relations between concepts to a rank (a measure) according to specific metrics such as pertinence to a specialty, quality of test, or severity of consequence. Given that reification is used in the SKOS model, we carried that over to our first OWL model. Returning to our example, we know that the ranking between (`malignantMelanomaOfBreast semrel:hasTreatmentProcedure radiationTherapy`), has a value of 6.0. We model this in OWL by introducing a new object property *hasRankedRelation* and the data property *hasRank* as follows:

`malignantMelanomaOfBreast` $\sqsubseteq$ `∃hasRankedRelation.`
      `(∃hasTreatmentProcedure.radiationTherapy ⊓ ∃hasRank.6)`

This reads as every instance of the class *malignantMelanomaOfBreast* has a *ranked relation* to an instance that both *has a treatment procedure that is radiationTherapy* and that *has rank 6*.

Meta data. All meta data, including alternate labels, preferred labels, and mappings to other vocabularies or ontologies were included as non-logical assertions, namely `owl:annotation` assertions, and also using `rdfs:labels` where appropriate.

All of these transformations are performed by a simple program written in Java and using the OWL API.

Results. The OWL lift resulted in an ontology with 927,941 classes with over 1.6 million asserted logical axioms. There were 45 object properties and one data property. The number of inferred atomic subsumptions stood at over 21 million. There were several issues that were revealed. The first issue was that there were numerous modelling errors from mapping narrower and broader relations to subclass and superclass relations. For example, consider the classes `Abortion` and `Abortion Recovery`. It is clear that `Abortion` is a broader term than `Abortion Recovery`, hence the use of a `skos:broader` relation in EMMeT. However, to state that one is a subclass of the other is just wrong: abortion recovery is not a kind of abortion. Thus we are at high risk of generating *incorrect* keys. That is, our technique could easily generate a multiple choice question where "abortion recovery" would be treated as a correct answer when it obviously is not.

The second issue was in relation to inheritance of superclass properties. Consider the disorder `Acute left ventricular failure`. According to EMMeT, this disorder is related via `has diagnostic procedure` to the class `echocardiography`. The relation has a rank of 6.0. A broader term for `Acute left ventricular failure` has the same procedure relation to `echocardiography` but with a rank of 10.0, which will be inherited into `Acute left ventricular failure` due to a subclass relation; i.e. it will have two rankings. This is clearly incorrect.

3.2 A More Sophisticated Approach

Of these two issues, the first is more challenging. The ranking issue can be handled simply by using an alternative modelling construct that avoids the problematic inheritance. We do not need to *interpret* or *validate* any of the specific rankings, thus the fixing remains domain independent. This is not true for the broader-narrower relations. There, we need to know *which ones* actually represent subsumptions (as many of them obviously do).

Ranking. The key to the second issues is that ranks are actually a sort of extra logical feature of EMMeT. Such extra-logical features are standardly represented with annotations and left for downstream processors to handle. OWL 2 added the ability to annotate whole axioms so this seemed like a good fit. Using the same example from Sect. 3.(1), we now introduce and annotation property `hasRank` in the following way

```
(malignantMelanomaOfBreast ⊑
    ∃hasTreatmentProcedure.radiationTherapy): hasRank:6
```

Annotation properties are not inherited to subclasses, and although we may lose any logical inferences w.r.t the data property version, since the annotation is basic, simple processing of the ontology would allow us to easily get the same information back.

SNOMED Alignment. To address the first issue, we needed a source of domain knowledge. Manual inspection was not a feasible option due to the size of EMMeT and lack of available domain expertise. However, EMMeT does have semantic relations which associates EMMeT concepts to concepts in external ontologies. Critically, one especially well connected source is SNOMED-CT [15]. SNOMED-CT, the coding controlled vocabulary is backed by a richly axiomitised OWL ontology and a long held focus on modelling domain relations correctly (esp. in a US context, though there are various internationalised versions and extensions). That, plus the fact that it has extensive coverage of the clinical domain makes it an ideal source of "ground truth" for subsumptions.

We did first try our question generation technique using SNOMED-CT alone but found that it lacked many relations between concepts that were essential for the sorts of questions we need to generate. For example, `isClinicalFinding For`, `hasTreatmentProcedure`, and `hasDiagnosticProcedure` are critical for formulating diagnostic puzzles, and are prevalent in EMMeT but not in SNOMED.

Out of the 927,941 EMMeT-in-OWL concepts, 106,435 contained mapping relationships to equivalent SNOMED CT classes. We first decided to test the accuracy of narrower broader relations of classes with SNOMED IDs against the same classes in SNOMED to see if a subclass relation was also present. There were over 1.4 million inferred atomic subsumptions in EMMeT for which both subclass and superclass had SNOMED IDs (6 % of all atomic subsumptions in EMMeT). From that 1.4 million, over 1.08 million (75 %) occurred in SNOMED, leaving 355,880 (25 %) not present in SNOMED. While it would still be a challenging task to review that remaining 25 % (which are potentially a source of clinical knowledge beyond SNOMED), having the large number of SNOMED valid subsumptions is more than sufficient for current purposes. Finally, there were 487,382 atomic subsumptions that were not present in EMMeT and were present in SNOMED. It seems somewhat implausible that these are "intentional" misses, that is, that the EMMeT designers think that SNOMED got some subsumptions wrong. These are a potential source for enriching EMMeT and we are investigating that possibility.

We identified two methods of including SNOMED CT in EMMeT: 1. Importing SNOMED directly and aligning the equivalent concepts and 2. Adding the subclass relations to existing EMMeT concepts. We also opted to encode `skos:narrower` and `skos:broader` relations as owl object properties instead of subclass and superclass relations, keeping there original intended semantics in both alignments. We decided to implement both methods and evaluated them against each other by comparing additional entailments when computing the class hierarchy in each ontology. Surprisingly, other than the entailments SNOMED alone provides, both ontologies did not provide any additional entailments upon classifying. After closer inspection we found that this was due to how we modelled the axioms. All logical axioms in the ontology are either of the form $A \sqsubseteq B$ or $A \sqsubseteq \exists R.B$. The left hand side of each axiom is always a named class, and since we have no definitions (axioms using $\equiv$ instead of $\sqsubseteq$) no further entailments could be inferred. Unfortunately, simply strengthening the axioms to equivalences would lead to a number of bogus results. (E.g., different diseases may have a common cause, so if we made each equivalent to having that cause we would end up with distinct diseases being equivalent.) We are currently investigating alternatives.

4   Related Work

Hepp and de Bruijn (2007) [6] introduced the GenTax algorithm to transform a thesaurus into a "useful" ontology by creating two OWL classes for each concept in the original hierarchy: (1) a *Gen*eric class and (2) a *Tax*onomy class that can be used to build a subsumption-based hierarchy that is valid with respect to a particular context of interest. Human intervention is required to determine (1) the main notions of the intended context, (2) a preliminary classification that is valid in the intended context. The obvious drawback of this methodology is the increased size of the resulting ontology since each concept in the original hierarchy is transformed into two separate concepts. The SKOS2OWL[5] tool is an implementation of the GenTax algorithm which offers a script-based transformation of SKOS informal hierarchies to OWL ontologies (with limited human intervention).

The method introduced by Wielinga et al. [19] is one of the earliest non-naive[6] transformations of thesauri to ontologies. The method was extended in [18,20] to transform two thesauri, namely MeSH [12] and WordNet [11], to OWL ontologies.

---

[5] http://www.heppnetz.de/projects/skos2owl/.

[6] By naive, here, we mean simply converting the hierarchy in the thesaurus to a subsumption hierarchy, ignoring any possible invalid consequences, as has been done in [8,9].

The work presented in this paper is also related to ontology learning techniques in that the ultimate goal of such techniques is to build an ontology from existing knowledge sources. Our approach makes use of existing thesauri, in particular SKOS knowledge bases, to build the ontology. Such knowledge bases are valuable sources for ontology learning as they already contain some hierarchical ordering between concepts and they can be rich in semantic relations. A review of other existing ontology learning techniques is out of the scope of this paper. The interested reader is referred to [4].

5 Conclusion

Given a bit of luck, we were able to produce a prototype version of EMMeT with sufficiently reliable subsumption relations for at least experimental work and perhaps even production quality generation of multiple choice questions. Futhermore, we seem to have identified an "easy win" extension of EMMeT with additional SNOMED relations. The effort involved was but a few person weeks and did not require expensive domain expertise.

On the downside, many broader-narrower relations are still "mysterious" from an ontological perspective. We just do not have enough information to know whether any of the residual relations are subsumptions (or definite *non*-subsumptions). We feel this exposes a weakness in the SKOS relaxed approach to relation specification: It inhibits reuse. Essentially, *only* applications that work at the same level of underspecification (or weaker) can use EMMeT without extensive examination. Whereas, if all the relations were *more specified* (not necessarily *as subsumptions* but at least as subsumptions *when they were* subsumptions), then use across various applications would be straight.

We will continue to strengthen EMMeTs hierarchy. As well as SNOMED-CT, there are also symbolic links to other formalisms such as UMLS and ICD which can be used to provide more entailments between concepts. It will be interesting to know how many of the narrower/broader relations are present as subclass relations w.r.t these formalisms. We also are exploring OWL modelling that would safely exploit the semantic relations we are currently modelling as subsumptions of existential restrictions.

References

1. Baader, F., Calvanese, D., McGuinness, D.L., Nardi, D., Patel-Schneider, P.F. (eds.): The Description Logic Handbook: Theory, Implementation and Applications, 2nd edn. Cambridge University Press, Cambridge (2007)
2. Bechhofer, S., Volz, R., Lord, P.: Cooking the semantic web with the OWL API. In: Fensel, D., Sycara, K., Mylopoulos, J. (eds.) ISWC 2003. LNCS, vol. 2870, pp. 659–675. Springer, Heidelberg (2003)
3. T. G. O. Consortium: Gene ontology: tool for the unification of biology. Nat. Genet. **25**(1), 25–29 (2000)

4. Drumond, L., Girardi, R.: A survey of ontology learning procedures. In: 3rd Workshop on Ontologies and Their Applications, vol. 427. CEUR Workshop Proceedings (2008)
5. Hedeler, C., Parsia, B., Brandt, S.: Estimating and analysing coordination in medical terminologies. In: IEEE 27th International Symposium on Computer-Based Medical Systems (CBMS), pp. 357–362 (2014)
6. Hepp, M., de Bruijn, J.: GenTax: a generic methodology for deriving OWL and RDF-S ontologies from hierarchical classifications, thesauri, and inconsistent taxonomies. In: Franconi, E., Kifer, M., May, W. (eds.) ESWC 2007. LNCS, vol. 4519, pp. 129–144. Springer, Heidelberg (2007)
7. Kazakov, Y., Krötzsch, M., Simancik, F.: Unchain my EL reasoner. In: Proceedings of the 24th International Workshop on Description Logics (DL-11) (2011)
8. Klein, M.: DAML+OIL and RDF Schema representation of UNSPSC (2015). http://www.cs.vu.nl/mcaklein/unspsc/. Accessed 4 Aug 2015
9. McGuinness, D.L.: UNSPSC Ontology in DAML+OIL (2015). http://www.ksl.stanford.edu/projects/DAML/UNSPSC.daml. Accessed 4 Aug 2015
10. Miles, A., Bechhofer, S.: SKOS simple knowledge organization system reference (2015). http://www.w3.org/TR/skos-reference/. Accessed 4 Aug 2015
11. Miller, G.: WordNet: a lexical database for English. Commun. ACM **38**(11) (1995)
12. Nelson, S.J.: Medical terminologies that work: the example of MeSH. In: Proceedings of the 10th International Symposium on Pervasive Systems, Algorithms, and Networks (ISPAN 2009), Kaohsiung, Taiwan, pp. 380–384, December 2009
13. Shearer, R., Motik, B., Horrocks, I.: HermiT: a highly-efficient OWL reasoner. In: Proceedings of the 5th International Workshop on OWL: Experiences and Directions (OWLED-08EU) (2008)
14. Sirin, E., Parsia, B., Cuenca Grau, B., Kalyanpur, A., Katz, Y.: Pellet: a practical OWL-DL reasoner. J. Web Semant. **5**(2), 51–53 (2007)
15. Spackman, K., Campbell, K.: Snomed CT: a reference terminology for health care. In: Masys, D.R. (ed.) Proceedings of AMIA Annual Fall Symposium, Bethesda, Maryland, USA, pp. 640–644. Hanley and Belfus Inc. (1997)
16. Stearns, M.Q., Price, C., Spackman, K.A., Wang, A.Y.: Snomed clinical terms: overview of the development process and project status. In: Proceedings of the 2001 AMIA Annual Symposium, pp. 662–666. Hanley and Belfus (2001)
17. Tsarkov, D., Horrocks, I.: FaCT++ description logic reasoner: system description. In: Proceedings of the 3rd International Joint Conference on Automated Reasoning (IJCAR) (2006)
18. van Assem, M., Menken, M.R., Schreiber, G., Wielemaker, J., Wielinga, B.J.: A method for converting thesauri to RDF/OWL. In: McIlraith, S.A., Plexousakis, D., van Harmelen, F. (eds.) ISWC 2004. LNCS, vol. 3298, pp. 17–31. Springer, Heidelberg (2004)
19. Wielinga, B.J., Schreiber, A.T., Sandberg, J.A.C.: From thesaurus to ontology. In: Proceedings of the First International Conference on Knowledge Capture (K-CAP 2001), Victoria, British Columbia, Canada (2001)
20. Wielinga, B.J., Wielemaker, J., Schreiber, G., van Assem, M.: Methods for porting resources to the semantic web. In: Bussler, C.J., Davies, J., Fensel, D., Studer, R. (eds.) ESWS 2004. LNCS, vol. 3053, pp. 299–311. Springer, Heidelberg (2004)

Experiences with Aber-OWL, an Ontology Repository with OWL EL Reasoning

Luke Slater[1], Miguel Ángel Rodríguez-García[1(✉)], Keiron O'Shea[1,2],
Paul N. Schofield[3], Georgios V. Gkoutos[2], and Robert Hoehndorf[1]

[1] Computational Bioscience Research Center, King Abdullah University
of Science and Technology, Thuwal 23955-6900, Kingdom of Saudi Arabia
{luke.slater,miguel.rodriguezgarcia,robert.hoehndorf}@kaust.edu.sa
[2] Aberystwyth University, Aberystwyth SY23 3DB, Wales, UK
keo7@aber.ac.uk
[3] University of Cambridge, Downing Street, Cambridge CB2 3EG, England, UK
pns12@hermes.cam.ac.uk

Abstract. Ontologies are widely used in biology and biomedicine for the annotation and integration of data, and hundreds of ontologies have been developed for this purpose. These ontologies also constitute large volumes of formalized domain knowledge, usually expressed in the Web Ontology Language (OWL). Computational access to the knowledge contained within them relies on the use of automated reasoning. We have developed Aber-OWL, an ontology repository that provides OWL EL reasoning to answer queries and verify the consistency of ontologies. Aber-OWL also provides a set of web services which provide ontology-based access to scientific literature in Pubmed and Pubmed Central, SPARQL query expansion to retrieve linked data, and integration with Bio2RDF. Here, we report on our experiences with Aber-OWL and outline a roadmap for future development. Aber-OWL is freely available at http://aber-owl.net.

Keywords: Biomedical ontology · Semantic web · Literature search · Semantic indexing · Query expansion

1 Introduction

Ontologies are used in most biological databases for the annotation and integration of data, and hundreds of ontologies have been developed for that purpose. These ontologies are commonly expressed in either the Web Ontology Language (OWL) [6] or an OWL-compatible language such as the OBO Flatfile Format [12]. Ontology repositories, such as BioPortal [14], the Ontology Lookup Service (OLS) [5] and OntoBee [18], currently provide web services and interfaces to access ontologies and their data in the biological domain. However, they do not utilize reasoning in the services they provide, and thus do not provide the advantages of semantic access, access to inferred knowledge and consistency verification.

© Springer International Publishing Switzerland 2016
V. Tamma et al. (Eds.): OWLED 2015, LNCS 9557, pp. 81–86, 2016.
DOI: 10.1007/978-3-319-33245-1_8

To enable this, we have created Aber-OWL [10] – an ontology repository in which access to ontologies is underpinned by reasoning. Aber-OWL consists primarily of an API, a web repository and a set of web services that provide ontology-based access to biological and biomedical data and literature. Here, we discuss our experiences with developing an ontology portal based on automated reasoning, discuss the current limitations, and suggest future extensions.

2   An Overview of Aber-OWL

2.1   Reasoning Services

The main component of Aber-OWL is a server that provides access to a large set of ontologies (currently 391) through an OWL EL reasoner. Ontologies are classified at the beginning of the server's runtime, and then kept in memory. We use the ELK reasoner [13], which supports the OWL EL profile, and any axioms that do not fall within the OWL EL subset are ignored. The restriction to OWL EL expressivity ensures that classification and query times remain tractable.

Access to the classified ontologies is provided through a REST API. This API can be utilized to perform Description Logic queries; specifically, it can be used to retrieve sub-, super-, or equivalent classes of a class description (which must also fall in the OWL EL profile). Querying is performed by transforming a class description in Manchester OWL Syntax [11] into an OWL class expression using the OWL API. If this transformation fails (e.g., when the query string provided is not a valid OWL class expression within the ontology being queried), an empty set of results is returned.

If the transformation succeeds, the ELK reasoner is used to retrieve sub-, super- or equivalent classes of the OWL class expression. Each query can be performed over a single or multiple ontologies stored within Aber-OWL. Consequently, results may be returned from multiple different ontologies at once. If a URL is specified as part of a query but the ontology is not available within Aber-OWL's repository, an attempt is made to retrieve the ontology from the URL, classify the ontology, perform the query over this ontology and return the results automatically. The API also provides additional ways to access the content of the ontologies, such as a substring-based search for classes, retrieving class descriptions based on the class IRI, and others.

2.2   Ontology-Based Data Access

One of our main aims in developing Aber-OWL is to demonstrate the potential for ontology-based access [4] to biological and biomedical data. Therefore, we developed several webservices that make use of Aber-OWL and combine OWL EL reasoning with access to different types of data sources.

The Aber-OWL: PubMed service is built on top of the Aber-OWL reasoning infrastructure, and retrieves articles in PubMed and PubMed Central in which any of the labels and synonyms of classes returned by a given semantic query appear. The literature search is performed over an Apache Lucene index holding all full text articles in PubMed Central and all abstracts in PubMed (using a disjunctive Lucene query of the class labels in the result set of the Aber-OWL query). This service allows, for example, to retrieve all articles that mention a subclass of `part-of some Heart` in its text.

The Aber-OWL: SPARQL service performs query expansion on a SPARQL query to incorporate the results of an Aber-OWL query. In particular, the set of class IRIs returned by an Aber-OWL query can be bound to a variable in SPARQL (using the SPARQL 1.1 `VALUES` statement) or used as an RDF collection that could, for example, be used with the `IN` operator as part of a `FILTER` statement. The use of Aber-OWL: SPARQL allows, for example, to query the UniProt [16] SPARQL endpoint for all proteins that have as their function a part of apoptosis that also regulates apoptosis (`part-of some 'apoptotic process' and regulates some 'apoptotic process'`). We further incorporated direct access to Bio2RDF [3] based on either the IRI of a class returned by an Aber-OWL query or based on the label of the class.

3   Experiences

One of the main challenges in developing an ontology portal based on OWL reasoning is the usability. Our target audience for Aber-OWL is twofold: on one side, we aim to provide services to bioinformaticians and ontologists who wish to make use of automated reasoning over ontologies as part of their workflow, and on the other hand, we aim to provide a useful repository of ontologies for biologists and biomedical researchers. While the first group of users will primarily use the API provided by Aber-OWL, the second group would rely mainly on the user interfaces we provide. However, making Description Logic querying easily accessible to a wide range of users through a common user interface is challenging and has constituted the main criticism we have received so far. To address these challenges in the future, we are considering utilizing natural language query interfaces [17], or visual construction of DL queries.

A related challenge is the automatic identification of labels and descriptions in ontologies. Across the range of over 390 ontologies in Aber-OWL, several different annotation properties are used to characterize the labels and textual descriptions of classes and object properties. Since Manchester OWL syntax relies on identifying natural language labels for classes so that they can be used as part of a class description, it is crucial to find a unified way of identifying labels, synonyms and descriptions of classes. The annotation properties we currently use to identify these are shown in Table 1, and they cover most of the ontologies in Aber-OWL. With the broad range of ontologies in Aber-OWL, the annotation properties in use will have to be constantly updated. As an intermediate solution, we now allow queries to be submitted in two forms, using either the labels of the classes and object properties, or using their IRIs directly.

With a more widespread adoption of Aber-OWL, we also have the potential for collecting a large set of real world Description Logic queries together with

Table 1. Labels, Synonyms and Descriptions used in Aber-OWL

Labels:
`rdfs:label`
http://www.w3.org/2004/02/skos/core#prefLabel
http://purl.obolibrary.org/obo/IAO_0000111
Synonyms:
http://www.w3.org/2004/02/skos/core#altLabel
http://purl.obolibrary.org/obo/IAO_0000118
http://www.geneontology.org/formats/oboInOwl#hasExactSynonym
http://www.geneontology.org/formats/oboInOwl#hasSynonym
http://www.geneontology.org/formats/oboInOwl#hasNarrowSynonym
http://www.geneontology.org/formats/oboInOwl#hasBroadSynonym
Descriptions:
http://purl.obolibrary.org/obo/IAO_0000115
http://www.w3.org/2004/02/skos/core#definition
http://purl.org/dc/elements/1.1/description
http://www.geneontology.org/formats/oboInOwl#hasDefinition

their execution time, which may become a useful resource for Description Logic reasoner performance evaluation [2]. We have created a log of all Description Logic queries submitted to Aber-OWL available at http://aber-owl.net/queries. log. The query log contains the ontology that has been queried, the kind of query made (retrieving sub-, super- or equivalent classes), the number of classes returned, and the time it took to execute the query.

4 Future Directions

In the future, we aim to further develop Aber-OWL in two major directions. First, we intend to explore how much of the semantics of ontologies can be made available in real time through an ontology portal. Currently, in Aber-OWL, we are using the ELK reasoner [13]. However, a large number of highly optimized reasoners are available, including some for more expressive fragments of OWL. We intend to evaluate some of these reasoners, based on the results achieved in the OWL Reasoner Evaluation challenges [2]. However, the theoretical limitations of non-tractable reasoning in OWL will remain a challenge, in particular with user-defined queries which may result in query times becoming too high. One solution to avoid this pitfall with more expressive fragments of OWL (or complete OWL 2) could be to set an upper limit for query answer time and fail if a query cannot be answered in that time, essentially resulting in incomplete reasoning. Nevertheless, such an approach could work if the majority of queries can be answered quickly.

Our second main aim for future development is to demonstrate additional functionality and novel types of bioinformatics applications that make use of inferences over ontologies. As our intended users fall in two categories (bioinformaticians/ontologists and domain experts), this step also takes two directions. For ontology developers in the biomedical domain, it is often difficult to evaluate the consequences of a change made to an ontology, since the ontology may be imported in multiple other ontologies. For example, a single change in the Gene Ontology [1], which is imported by a large number of other ontologies, can have a significant impact on any of the other ontologies, such as resulting in incoherent class definitions or leading to inconsistency. At the moment, such consequences are not visible to the ontology developers. Aber-OWL has the potential of immediately showing the consequences of such a change across the range of ontologies it contains, essentially serving as a continuous integration environment for distributed development of ontologies.

Our other target audience, the domain experts, often work with ontologies as graph structures [9] that are used in visualization and data analysis. We intend to generate and visualize ontology graph structures, including the graph structures induced by axiom patterns [7], in addition to the subsumption hierarchy currently available through Aber-OWL.

5   Conclusion

Despite Aber-OWL being relatively new, we have already established a small user base, mainly for the REST API services. We have also demonstrated that reasoning even over a large set of ontologies is now a possibility and can be performed efficiently [15], and that novel kinds of applications can be developed which rely on automated reasoning and semantic query. These applications may even lead to new data- or text-mining methods that reveal new insights into a domain of knowledge [8]. In the future, we hope that Aber-OWL will establish itself as an ontology repository in the biological and biomedical domain that makes the semantics of ontologies and inferences over them available to a wide range of users.

References

1. Ashburner, M., Ball, C.A., Blake, J.A., Botstein, D., Butler, H., Cherry, M.J., Davis, A.P., Dolinski, K., Dwight, S.S., Eppig, J.T., Harris, M.A., Hill, D.P., Tarver, L.I., Kasarskis, A., Lewis, S., Matese, J.C., Richardson, J.E., Ringwald, M., Rubin, G.M., Sherlock, G.: Gene ontology: tool for the unification of biology. Nat. Genet. **25**(1), 25–29 (2000). http://dx.doi.org/10.1038/75556
2. Bail, S., Glimm, B., Jiménez-Ruiz, E., Matentzoglu, N., Parsia, B., Steigmiller, A. (eds.): ORE 2014: OWL Reasoner Evaluation Workshop. No. 1207 in CEUR Workshop Proceedings. CEUR-WS.org, Aachen, Germany (2014)
3. Belleau, F., Nolin, M., Tourigny, N., Rigault, P., Morissette, J.: Bio2RDF: towards a mashup to build bioinformatics knowledge systems. J. Biomed. Inform. **41**(5), 706–716 (2008). http://dx.doi.org/10.1016/j.jbi.2008.03.004
4. Bienvenu, M., ten Cate, B., Lutz, C., Wolter, F.: Ontology-based data access: a study through disjunctive datalog, CSP, and MMSNP. In: PODS, pp. 213–224 (2013)

5. Cote, R., Jones, P., Apweiler, R., Hermjakob, H.: The ontology lookup service, a lightweight cross-platform tool for controlled vocabulary queries. BMC Bioinform. **7**(1), 97 (2006). http://dx.doi.org/10.1186/1471-2105-7-97

6. Grau, B., Horrocks, I., Motik, B., Parsia, B., Patelschneider, P., Sattler, U.: OWL 2: the next step for OWL. Web Seman. Sci. Serv. Agents World Wide Web **6**(4), 309–322 (2008). http://dx.doi.org/10.1016/j.websem.2008.05.001

7. Hoehndorf, R., Oellrich, A., Dumontier, M., Kelso, J., Rebholz-Schuhmann, D., Herre, H.: Relations as patterns: bridging the gap between OBO and OWL. BMC Bioinform. **11**(1), 441 (2010)

8. Hoehndorf, R., Schofield, P.N., Gkoutos, G.V.: Analysis of the human diseasome using phenotype similarity between common, genetic, and infectious diseases. Sci. Rep. **5**, Article no. 10888 (2015). http://bmcbioinformatics.biomedcentral.com/articles/10.1186/1471-2105-11-441

9. Hoehndorf, R., Schofield, P.N., Gkoutos, G.V.: The role of ontologies in biological and biomedical research: a functional perspective. Briefings Bioinform. **16**(6), 1069–1080 (2015). https://bib.oxfordjournals.org/content/16/6/1069.full

10. Hoehndorf, R., Slater, L., Schofield, P.N., Gkoutos, G.V.: Aber-OWL: a framework for ontology-based data access in biology. BMC Bioinform. **16**, 26 (2015). http://www.biomedcentral.com/1471-2105/16/26/abstract

11. Horridge, M., Drummond, N., Goodwin, J., Rector, A., Stevens, R., Wang, H.: The Manchester OWL syntax. In: Proceedings of the 2006 OWL Experiences and Directions Workshop (OWL-ED 2006) (2006)

12. Horrocks, I.: OBO flat file format syntax and semantics and mapping to OWL Web Ontology Language. Technical report, University of Manchester, March 2007. http://www.cs.man.ac.uk/~horrocks/obo/

13. Kazakov, Y., Krötzsch, M., Simancik, F.: The incredible ELK. J. Autom. Reasoning **53**(1), 1–61 (2014). http://dx.doi.org/10.1007/s10817-013-9296-3

14. Noy, N.F., Shah, N.H., Whetzel, P.L., Dai, B., Dorf, M., Griffith, N., Jonquet, C., Rubin, D.L., Storey, M.A.A., Chute, C.G., Musen, M.A.: Bioportal: ontologies and integrated data resources at the click of a mouse. Nucleic Acids Res. **37**(Web Server issue), W170–W173 (2009). http://dx.doi.org/10.1093/nar/gkp440

15. Slater, L., Gkoutos, G., Schofield, P.N., Hoehndorf, R.: Using Aber-OWL for fast and scalable reasoning over BioPortal ontologies. In: Proceedings of International Conference on Biomedical Ontologies (ICBO), pp. 72–76, July 2015. http://icbo2015.fc.ul.pt/ICBO2015Proceedings.pdf

16. The Uniprot Consortium: The universal protein resource (uniprot). Nucleic Acids Res 35(Database issue), January 2007. http://view.ncbi.nlm.nih.gov/pubmed/17142230

17. Usbeck, R., Ngonga Ngomo, A.C., Bühmann, L., Unger, C.: HAWK - hybrid question answering over linked data. In: 12th Extended Semantic Web Conference, 31st May – 4th 2015, Portoroz, Slovenia, June 2015

18. Xiang, Z., Mungall, C.J., Ruttenberg, A., He, Y.: Ontobee: a linked data server and browser for ontology terms. In: Proceedings of International Conference on Biomedical Ontology, pp. 279–281 (2011)

Towards a Rule Based Distributed OWL Reasoning Framework

Raghava Mutharaju[✉], Prabhaker Mateti, and Pascal Hitzler

Wright State University, Dayton, OH, USA
{mutharaju.2,prabhaker.mateti,pascal.hitzler}@wright.edu

Abstract. The amount of data exposed in the form of RDF and OWL continues to increase exponentially. Some approaches have already been proposed for the scalable reasoning over several language profiles such as RDFS, OWL Horst, OWL 2 EL, OWL 2 RL etc. But all those approaches are limited to the particular ruleset that the reasoner supports. In this work, we propose the idea for a rule-based distributed reasoning framework that can support any given ruleset and highlight some of the challenges that needs to be solved in order to implement such a framework.

1 Introduction

The W3C recommendations RDF and OWL, are primarily used to represent data in the Semantic Web. Large amount of data in these formats is now available and it only continues to grow. Several billions of RDF triples are available as Linked Open Data (close to 90 billion[1]). Automated generation of OWL axioms from streaming data [9] and text [10] can result in very large knowledge bases. Reasoning is one of the most important operations that can be performed over OWL and RDF knowledge bases. It is required to infer logical consequences and to check the consistency of the knowledge base. Reasoning is memory and compute intensive. So reasoning over large knowledge bases needs a scalable approach. Currently, all the popular off-the-shelf reasoners work only on a single machine, possibly with multiple cores. It is not possible for a single machine to keep up with the growth rate of data. Also, for some reasoning tasks the output is several times larger than the input. Distributed memory reasoning provides a viable alternative.

There are some existing approaches for scalable reasoning over each individual Semantic Web language profile such as RDFS, OWL Horst, OWL 2 EL, OWL 2 RL (see Sect. 4). Reasoning over ontologies in each of these profiles is performed using a set of rules that vary with each profile (there is some overlap among the different rulesets). Generally, the existing solutions are tuned towards a particular ruleset and are not adaptable to other rulesets. This poses a problem for users who work with multiple rulesets and also in cases where users need a scalable solution for a ruleset which does not have a customized approach. In this paper, we propose the idea for a unified distributed reasoning framework that

---

[1] http://stats.lod2.eu/.

© Springer International Publishing Switzerland 2016
V. Tamma et al. (Eds.): OWLED 2015, LNCS 9557, pp. 87–92, 2016.
DOI: 10.1007/978-3-319-33245-1_9

Table 1. RDFS closure rules

1: $s\ p\ o$ (if o is literal)	$\Rightarrow$ _:n rdf:type rdfs:Literal
2: p rdfs:domain x & $s\ p\ o$	$\Rightarrow$ s rdf:type x
3: p rdfs:range x & $s\ p\ o$	$\Rightarrow$ o rdf:type x
4a: $s\ p\ o$	$\Rightarrow$ s rdf:type rdfs:Resource
4b: $s\ p\ o$	$\Rightarrow$ o rdf:type rdfs:Resource
5: p rdfs:subPropertyOf q & q rdfs:subPropertyOf r	$\Rightarrow$ p rdfs:subPropertyOf r
6: p rdf:type rdf:Property	$\Rightarrow$ p rdfs:subPropertyOf p
7: $s\ p\ o$ & p rdfs:subPropertyOf q	$\Rightarrow$ $s\ q\ o$
8: s rdf:type rdfs:Class	$\Rightarrow$ s rdfs:subClassOf rdfs:Resource
9: s rdf:type x & x rdfs:subClassOf y	$\Rightarrow$ s rdf:type y
10: s rdf:type rdfs:Class	$\Rightarrow$ s rdfs:subClassOf s
11: x rdfs:subClassOf y & y rdfs:subClassof z	$\Rightarrow$ x rdfs:subClassOf z
12: p rdf:type rdfs:ContainerMembershipProperty	$\Rightarrow$ p rdfs:subPropertyOf rdfs:member
13: o rdf:type rdfs:Datatype	$\Rightarrow$ o rdfs:subClassOf rdfs:Literal

can work on any given ruleset. This framework can, not only handle the afore-mentioned language profiles but also avoids the need to develop a customized scalable approach for any new ruleset.

2   Challenges

Some of the challenges in the design and implementation of a rule-based distributed reasoning framework are discussed here.

2.1   Rule Dependency Analysis

If the input to a rule R_i, depends on the output of another rule R_j, then the rule R_i is dependent on R_j. The more independent the rules are, better can be the rule distribution among the nodes in the cluster.

A rule dependency graph can be constructed in order to determine the inter-dependency among the rules. Each vertex represents a rule and an outgoing edge between vertex v_i and v_j represents the dependency of vertex v_i on v_j. Isolated vertices are independent of each other and can be executed in parallel. For each vertex v_i, all the vertices that are reachable from it are dependent on each other.

RDFS rules from [21] are shown in Table 1 and its dependency graph is shown in Fig. 1. There are several rules that are independent and can be given to separate nodes in the cluster. Rules 5, 6 and 7 are dependent on each other and can be grouped together i.e., all the three rules can be executed by one node. Same holds for the group of rules 9, 2, 3, 10, 11.

On the other hand, rules for $\mathcal{EL}^{++}$ [1] are highly inter-dependent and are difficult to parallelize.

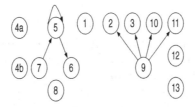

Fig. 1. RDFS rule dependency graph

2.2 Data Distribution

Rule dependency analysis can provide the basis for data distribution. Some rules are applicable only to a specific type of data and in these cases, it would be easy to partition the data based on the rule distribution. Cases such as rules 4a, 4b and 6 in Table 1 should be handled differently since they are applicable to the entire dataset. These rules should be run on all the nodes of the cluster so as to avoid data overloading. Heuristics such as number of variables in a rule in proportion to the constants could be used to determine such type of rules.

2.3 Rule Implementation

Interpreting different rulesets will be very difficult for the framework. Instead, rulesets should be converted to a common domain specific language (DSL) that is supported by the framework. This DSL should be able to define the vocabulary, syntax and semantics of the language to be used.

For the choice of DSL, there are some options. (1) general purpose rule languages such as RETE, Datalog and Prolog. (2) or a custom DSL for the rules supported by the framework. DSL should support the declaration of variables and constants in the rules.

3 Evaluation Plan

The rule-based distributed reasoning framework can be evaluated along the lines of adaptation and extension.

- There are several existing specialized and scalable reasoners for rulesets such as RDFS (WebPIE, Cichlid), OWL Horst (QueryPIE) and OWL 2 EL (DistEL). The framework should be able to handle these rulesets. The performance of the general purpose framework in comparison to the specialized ones remains to be seen.
- The framework should be able to take in a new ruleset and provide sound and complete inference over the given data.

4   Related Work

There are several language profiles in the Semantic Web that support rule-based reasoning. Other reasoning approaches such as tableau algorithms are not considered here. Scalable reasoning approaches such as parallel shared memory reasoning to distributed shared-nothing reasoning exist.

RDFS has around 34 inference (entailment) rules including simple, extensional and datatype entailment rules [3]. Almost all of the existing work on scalable RDFS entailment (closure) computation considers only a subset of these rules. Marvin computes the closure of RDF triples using a peer-to-peer model [16]. In [22], triples are distributed across the cluster by making a distinction between the schema and instance triples. Finding closure becomes an embarrassingly parallel computation. Several other scalable approaches exist for RDFS closure computation [4,6,21].

OWL Horst (also known as pD*) [5] extends RDFS entailment rules to include reasoning with datatypes from a given datatype map D. Rule partitioning and data partitioning strategies are explored in [19] for computing closure over OWL Horst knowledge base. QueryPIE [20] is a backward chaining distributed reasoner that supports OWL Horst reasoning over large knowledge bases. A recent Apache Spark implementation of RDFS and OWL Horst rules named Cichlid, is 10 times faster than state-of-the-art implementations [2].

OWL 2 RL [11] is further extension of pD* and has several entailment rules. A scalable OWL 2 RL inference engine has been implemented inside a relational database system (Oracle) in [8]. QueryPIE has partial support for OWL 2 RL.

The description logic underlying OWL 2 EL is $\mathcal{EL}^{++}$ and there are 11 completion rules ([1]) for classification (computing the subsumption hierarchy of all concepts). Three approaches to distributed $\mathcal{EL}^{++}$ reasoning were discussed in [13] including MapReduce [15]. Among them, the most efficient system named DistEL, follows a peer-to-peer model that uses rule partitioning based on the axiom types [14]. Though not distributed, parallelization of OWL 2 EL classification has been studied in [7,18].

There are some existing generic rule-based scalable reasoning approaches. RETE implementation on GPUs for RDFS and OWL Horst rulesets is shown in [17]. Another alternative is to convert different rulesets into datalog rules. A parallel implementation of datalog programs with application to RDFS rules is shown in [12]. A distributed approach for either of these two has not been developed yet.

5   Conclusion

There are different rulesets for different language profiles and there are scalable approaches for many of these rulesets. However, such a specialized scalable reasoner does not work on other rulesets. A general purpose rule-based distributed reasoning framework is proposed here to fill that gap. This framework provides more flexibility in terms of rulesets but there could be a possible loss in performance when compared to specialized scalable reasoners. A proper evaluation of the framework is required in order to determine its flexibility and performance.

The next step is to choose the appropriate DSL by checking the advantages and disadvantages of RETE, Datalog and Prolog. After this, we plan to proceed with the implementation of rest of the pieces in the framework.

References

1. Baader, F., Brandt, S., Lutz, C.: Pushing the EL envelope. In: Kaelbling, L.P., Saffiotti, A. (eds.) Proceedings of the Nineteenth International Joint Conference on Artificial Intelligence, IJCAI-2005, July 30–August 5 2005, Edinburgh, Scotland, UK, pp. 364–369. AAAI (2005)
2. Gu, R., Wang, S., Wang, F., Yuan, C., Huang, Y.: Cichlid: efficient large scale RDFS/OWL reasoning with spark. In: 2015 IEEE International Parallel and Distributed Processing Symposium, IPDPS 2015, 25–29 May 2015, Hyderabad, India, pp. 700–709. IEEE Computer Society (2015)
3. Hayes, P., Patel-Schneider, P.F.: RDF Semantics (2014). http://www.w3.org/TR/rdf11-mt/
4. Heino, N., Pan, J.Z.: RDFS reasoning on massively parallel hardware. In: Cudré-Mauroux, P., et al. (eds.) ISWC 2012, Part I. LNCS, vol. 7649, pp. 133–148. Springer, Heidelberg (2012)
5. ter Horst, H.J.: Combining RDF and part of OWL with rules: semantics, decidability, complexity. In: Gil, Y., Motta, E., Benjamins, V.R., Musen, M.A. (eds.) ISWC 2005. LNCS, vol. 3729, pp. 668–684. Springer, Heidelberg (2005)
6. Kaoudi, Z., Miliaraki, I., Koubarakis, M.: RDFS reasoning and query answering on top of DHTs. In: Sheth, A.P., Staab, S., Dean, M., Paolucci, M., Maynard, D., Finin, T., Thirunarayan, K. (eds.) ISWC 2008. LNCS, vol. 5318, pp. 499–516. Springer, Heidelberg (2008)
7. Kazakov, Y., Krötzsch, M., Simančík, F.: Concurrent classification of $\mathcal{EL}$ ontologies. In: Aroyo, L., Welty, C., Alani, H., Taylor, J., Bernstein, A., Kagal, L., Noy, N., Blomqvist, E. (eds.) ISWC 2011, Part I. LNCS, vol. 7031, pp. 305–320. Springer, Heidelberg (2011)
8. Kolovski, V., Wu, Z., Eadon, G.: Optimizing enterprise-scale OWL 2 RL reasoning in a relational database system. In: Patel-Schneider, P.F., Pan, Y., Hitzler, P., Mika, P., Zhang, L., Pan, J.Z., Horrocks, I., Glimm, B. (eds.) ISWC 2010, Part I. LNCS, vol. 6496, pp. 436–452. Springer, Heidelberg (2010)
9. Lécué, F., Tucker, R., Bicer, V., Tommasi, P., Tallevi-Diotallevi, S., Sbodio, M.: Predicting severity of road traffic congestion using semantic web technologies. In: Presutti, V., d'Amato, C., Gandon, F., d'Aquin, M., Staab, S., Tordai, A. (eds.) ESWC 2014. LNCS, vol. 8465, pp. 611–627. Springer, Heidelberg (2014)
10. Lehmann, J.: DL-learner: learning concepts in description logics. J. Mach. Learn. Res. (JMLR) **10**, 2639–2642 (2009)
11. Motik, B., Grau, B.C., Horrocks, I., Wu, Z., Fokoue, A., Lutz, C. (eds.): OWL 2 Web Ontology Language Profiles. In: W3C Recommendation (2012). http://www.w3.org/TR/owl2-profiles/
12. Motik, B., Nenov, Y., Piro, R., Horrocks, I.: Parallel materialisation of datalog programs in main-memory RDF databases. In: Brodley, C.E., Stone, P. (eds.) Proceedings of the Twenty-Eighth AAAI Conference on Artificial Intelligence, July 27–31 2014, Qébec City, Qébec, Canada. AAAI Press (2014)

13. Mutharaju, R., Hitzler, P., Mateti, P.: Distributed OWL EL reasoning: the story so far. In: Liebig, T., Fokoue, A. (eds.) Proceedings of the 10th International Workshop on Scalable Semantic Web Knowledge Base Systems, Riva Del Garda, Italy. CEUR Workshop Proceedings, vol. 1261, pp. 61–76. CEUR-WS.org (2014)
14. Mutharaju, R., Hitzler, P., Mateti, P., Lécué, F.: Distributed and Scalable OWL EL Reasoning. In: Gandon, F., Sabou, M., Sack, H., d'Amato, C., Cudré-Mauroux, P., Zimmermann, A. (eds.) ESWC 2015. LNCS, vol. 9088, pp. 88–103. Springer, Heidelberg (2015)
15. Mutharaju, R., Maier, F., Hitzler, P.: A MapReduce algorithm for EL+. In: Haarslev, V., Toman, D., Weddell, G.E. (eds.) Proceedings of the 23rd International Workshop on Description Logics (DL 2010), 4–7 May 2010, Waterloo, Ontario, Canada. CEUR Workshop Proceedings, vol. 573. CEUR-WS.org (2010)
16. Oren, E., Kotoulas, S., Anadiotis, G., Siebes, R., ten Teije, A., van Harmelen, F.: Marvin: distributed reasoning over large-scale semantic web data. Web Seman. Sci. Serv. Agents World Wide Web 7(4), 305–316 (2009)
17. Peters, M., Sachweh, S., Zündorf, A.: Large scale rule-based reasoning using a laptop. In: Gandon, F., Sabou, M., Sack, H., d'Amato, C., Cudré-Mauroux, P., Zimmermann, A. (eds.) ESWC 2015. LNCS, vol. 9088, pp. 104–118. Springer, Heidelberg (2015)
18. Ren, Y., Pan, J.Z., Lee, K.: Parallel ABox reasoning of $\mathcal{EL}$ ontologies. In: Pan, J.Z., Chen, H., Kim, H.-G., Li, J., Horrocks, I., Mizoguchi, R., Wu, Z., Wu, Z. (eds.) JIST 2011. LNCS, vol. 7185, pp. 17–32. Springer, Heidelberg (2012)
19. Soma, R., Prasanna, V.K.: Parallel inferencing for OWL knowledge bases. In: 2008 International Conference on Parallel Processing, ICPP 2008, 8–12 September 2008, Portland, Oregon, USA, pp. 75–82. IEEE Computer Society (2008)
20. Urbani, J., van Harmelen, F., Schlobach, S., Bal, H.: QueryPIE: backward reasoning for OWL horst over very large knowledge bases. In: Aroyo, L., Welty, C., Alani, H., Taylor, J., Bernstein, A., Kagal, L., Noy, N., Blomqvist, E. (eds.) ISWC 2011, Part I. LNCS, vol. 7031, pp. 730–745. Springer, Heidelberg (2011)
21. Urbani, J., Kotoulas, S., Oren, E., van Harmelen, F.: Scalable distributed reasoning using MapReduce. In: Bernstein, A., Karger, D.R., Heath, T., Feigenbaum, L., Maynard, D., Motta, E., Thirunarayan, K. (eds.) ISWC 2009. LNCS, vol. 5823, pp. 634–649. Springer, Heidelberg (2009)
22. Weaver, J., Hendler, J.A.: Parallel materialization of the finite RDFS closure for hundreds of millions of triples. In: Bernstein, A., Karger, D.R., Heath, T., Feigenbaum, L., Maynard, D., Motta, E., Thirunarayan, K. (eds.) ISWC 2009. LNCS, vol. 5823, pp. 682–697. Springer, Heidelberg (2009)

Improving OWL RL Reasoning in N3 by Using Specialized Rules

Dörthe Arndt[1(✉)], Ben De Meester[1], Pieter Bonte[2], Jeroen Schaballie[2],
Jabran Bhatti[3], Wim Dereuddre[3], Ruben Verborgh[1], Femke Ongenae[2],
Filip De Turck[2], Rik Van de Walle[1], and Erik Mannens[1]

[1] Ghent University – iMinds – Data Science Lab, Ghent, Belgium
{doerthe.arndt,ben.demeester,ruben.verborgh}@ugent.be
[2] IBCN Research Group, INTEC, Ghent University – iMinds, Ghent, Belgium
{pieter.bonte,jeroen.schaballie,femke.ongenae}@intec.ugent.be
[3] Televic Healthcare, Izegem, Belgium
{j.bhatti,w.dereuddre}@televic.com

Abstract. Semantic Web reasoning can be a complex task: depending
on the amount of data and the ontologies involved, traditional OWL DL
reasoners can be too slow to face problems in real time. An alternative
is to use a rule-based reasoner together with the OWL RL/RDF rules
as stated in the specification of the OWL 2 language profiles. In most
cases this approach actually improves reasoning times, but due to the
complexity of the rules, not as much as it could. In this paper we present
an improved strategy: based on the TBoxes of the ontologies involved
in a reasoning task, we create more specific rules which then can be
used for further reasoning. We make use of the EYE reasoner and its
logic Notation3. In this logic, rules can be employed to derive new rules
which makes the rule creation a reasoning step on its own. We evaluate
our implementation on a semantic nurse call system. Our results show
that adding a pre-reasoning step to produce specialized rules improves
reasoning times by around 75 %.

Keywords: Notation3 · Rule-based reasoning · OWL 2 RL

1 Introduction

With the increasing amount of carefully designed ontologies semantic web reasoning is becoming more popular for industrial applications: ontologies can be
employed to solve complex problems in domains like medicine, automotive industry or finance (e.g., [13,17]). Nevertheless, there are still some obstacles which
hinder semantic web reasoning from being fully established. One of these is scalability: depending on the amount of data and the ontologies involved, traditional
OWL DL reasoners can be too slow to solve problems in real time. The different
OWL 2 profiles [9] provide a solution: by using less expressive but still powerful
subsets of OWL DL, reasoning times can be significantly improved.

© Springer International Publishing Switzerland 2016
V. Tamma et al. (Eds.): OWLED 2015, LNCS 9557, pp. 93–104, 2016.
DOI: 10.1007/978-3-319-33245-1_10

In this paper we focus on the OWL RL profile which is designed to enable rule-based reasoners to draw the right conclusions from ontology data and concepts. The rules for that, as presented in the specification, are complex in the sense that they rely on rather complicated patterns occurring in both ABox and TBox which have to be found to draw conclusions. We propose to improve OWL RL reasoning performance by adding an extra reasoning step. Based on the ontology's TBox, specialized rules can be automatically produced to be used for further reasoning on the ABox. Due to its expressiveness we use Notation3 Logic [6] to perform this task. The highly performant EYE reasoner [19] is used for reasoning. As our pre-reasoning step has to be executed only once for every TBox, our approach is especially suitable for situations where the same reasoning has to be performed on frequently changing data. We tested our implementation in an event based reasoning set-up: a semantic nurse call system which controls the technical equipment in a hospital and, for example, assigns the most suitable nurse to a patient's call. Our tests showed that our pre-reasoning step reduces reasoning times at about 75 % compared to an implementation using the originally proposed rules.

The remainder of this paper is structured as follows: in Sect. 2 we give an overview of related work. After that, in Sect. 3, we explain our use case, a semantic nurse call system. Section 4 gives a general introduction to OWL RL in N3. In Sect. 5 we describe our system in more detail, focusing in particular on the improved rules themselves and the steps which are necessary to produce them. An evaluation of our implementation is given in Sect. 6. We summarize our main findings and give an outlook to future work in Sect. 7.

2   Related Work

Traditionally, reasoning over OWL ontologies was performed by description logic based reasoners using (variants of) the tableaux algorithm. Prominent examples of such reasoners are Pellet [16] and HermiT [18]. Both support—as others of their kind—the full OWL DL profile. The expressiveness of this profile and the complexity of the related reasoning, make these reasoners perform rather slow in comparison with, for example, rule-based reasoners. The OWL 2 profiles [9] aim to overcome this gap by defining less expressive but still powerful subsets of OWL DL. One of these profiles is OWL RL, which was designed to enable rule-based reasoners to cope with OWL ontologies. Various implementations make use of the OWL 2 RL/RDF rules as proposed in the specification, among them OWLim [8] and Oracle's RDF Semantic Graph [20]. As most other implementations we are aware of, these reasoners support their own rule format, and optimizations are done internally using the underlying programming language. We propose an optimization which can be done in the logic itself by performing an extra reasoning step. We are thereby independent of a specific reasoner.

Notation3 Logic (N_3) was introduced in 2008 by Tim Berners-Lee et al. [6]. It forms a superset of RDF and extends the RDF data model by formulas (graphs), functional predicates, universal variables and logical operators, in particular the implication operator. Rules in N_3 can not only be applied to derive new RDF triples, it is also possible to write and apply rules with new rules in their consequence, and thus to derive new rules. It is exactly this property which made us opt for using N_3 instead of other rule formats like, e.g., SWRL [14].

There are several reasoners supporting N_3: FuXi [1] is a forward-chaining production system for Notation3 whose reasoning is based on the RETE algorithm. The forward-chaining cwm [4] reasoner is a general-purpose data processing tool which can be used for querying, checking, transforming and filtering information. EYE [19] is a reasoner which is enhanced with Euler path detection. It supports backward and forward reasoning and also a user-defined mixture of both. Amongst its numerous features are the option to skolemize blank nodes and the possibility to produce and reuse proofs for further reasoning. The reason why we use EYE in our implementation is its high performance. Existing benchmarks and results are listed in the above-mentioned paper [19] and on the EYE website [11].

3 Use Case

Our use case is a nurse call system in a hospital. The system is aware of certain details about personnel and patients represented in an OWL ontology. Such information can include: personal skills of a staff member, staff competences, patient information, special patient needs, and/or the personal relationship between staff members and patients. Furthermore, there is dynamic information available, for example, the current location of staff members and their status (busy or free). When a call is made, the nurse call system should be able to assign the best staff member to answer that call. The definition of this "best" person varies between hospitals and can be quite complex. The system additionally controls different devices. If for example staff members enter a room with a patient, a light should be switched on; if they log into the room's terminal, they should have access to the medical lockers in the room. The event-driven reasoning system for this use case has to fulfill certain requirements.

scalability It should cope with data sets ranging from 1000 to 100,000 relevant triples (i.e., triples necessary to be included for the reasoning to be correct). Especially in bigger hospitals the number of staff members and patients and thereby also the amount of available information about those can be quite big. It is not always possible to divide this knowledge into smaller independent chunks as this data is normally full of mutual dependencies.

functional complexity It should implement deterministic decision trees with varying complexities. The reasons to assign a nurse to a certain patient can be as manifold as the data. Previous work has shown that this complexity is not only theoretically possible but also desired by the parties interested in such a semantic system [15].

configuration It should support the ability to change these decision trees at configuration time. Different hospitals have different requirements and even in one single hospital those requirements can easily change due to e.g., an

```
1   @prefix rdfs: <http://www.w3.org/2000/01/rdf-schema#>.
2   @prefix rdf: <http://www.w3.org/1999/02/22-rdf-syntax-ns#>.
3
4   {?C rdfs:subClassOf ?D. ?X a ?C} => {?X a ?D}.
```

Listing 1. OWL RL rule for `rdfs:subClassOf` class axiom in N3.

increase of available information or a simple change in the hospital's organizational concepts or philosophy.

real-time It should return a response within 5 s to any given event. Especially in such a delicate sector as patient care, seconds can make a difference. Even though a semantic nurse call system will not typically be employed to assign urgent emergency calls through complex decision trees, a patient should not wait too long till his possibly pressing request is answered.

The functional complexity requirement together with the configuration constraint motivate the choice of a reasoning system which supports rules as these can be seen as the most natural way to express decision trees. Even though a numerous amount of OWL DL reasoners support at least one rule format, their reasoning is too slow to meet the scalability and the real-time constraint [2]. Therefore, we chose a rule-based solution.

4   OWL RL in N$_3$

In a first attempt to solve the above-mentioned problem we used a direct translation of OWL 2 RL/RDF rules as listed on the corresponding website [9]. Where possible, we made use of existing N3-translations of these rules as provided by EYE [12]. Missing concepts were added. The data was represented using the ACCIO ontology [15] which will be further described in Sect. 6.1. The results of this implementation were already promising [2], but for larger data sets the reasoning took multiple minutes and, thus, did not meet the requirements claimed above.

We explain the idea behind these OWL RL rules in N$_3$ and how they can be improved using an example: Listing 1 shows the class axiom rule[1] which is needed to deal with the rdfs concept `subclassOf`. For convenience we omit the prefixes in the formulas below. The empty prefix refers to the ACCIO ontology, `rdf` and `rdfs` have the same meaning as in Listing 1. Consider that we have the following TBox triple stating that the class `:Call` is a subclass of the class `:Task`:

$$:Call\ rdfs:subClassOf\ :Task. \qquad (1)$$

If the ABox contains an individual which is member of the class `:Call`

$$:call1\ a\ :Call. \qquad (2)$$

---

[1] The rule is the N3 version of the cax-sco rule in Table 7 on the OWL 2 Profiles website [9].

an OWL DL reasoner would make the conclusion that the individual also belongs to the class `Task`:

$$:\texttt{call1 a :Task.} \tag{3}$$

Our rule in Listing 1 does exactly the same: as Formulas 1 and 2 can be unified with the antecedent of the rule, a reasoner derives the triple in Formula 3. But this unification is rather expensive: if we take a closer look to the antecedent we see that it contains three different variables occurring in two different triples which have to be instantiated with the data of the ontology. In our use case information as stated in Formula 2 can change—patients will make new calls—but statements as Formula 1 can be considered as fixed: the terminology does not change during the reasoning process, calls are tasks for our ontology. Our solution makes use of this observation: what is valid for the triple in Formula 1 also counts for other TBox-triples. We consider the TBox as static knowledge which can be used for pre-processing. The idea of our solution is to do as much unification as possible before dealing with (possibly) dynamic data. We produce more specialized rules, in the case mentioned above, for example the rule

$$\{\texttt{?X a :Call.}\} \Rightarrow \{\texttt{?X a :Task.}\}. \tag{4}$$

which will derive for every new call, that it is also a task, just as the rule in Listing 1 does.

5 Producing TBox-rules

In order to achieve the goal explained in the last section, producing specialized rules based on the concepts present in the ontology's TBox, we use the EYE reasoner. Reasoning in EYE can be considered as a single process, having as input all necessary files representing the knowledge (i.e., the necessary ontologies, data, and rule-files), and a query-file that filters the output of the reasoning result. We have to perform two steps:

1. Produce a grounded copy of the TBox.
2. Use rules to translate the grounded TBox into specialized rules.

The need of the first step has to do with the fact that an ontology can contain anonymous classes represented by blank nodes. Used in rules, these blank node class names have, due to the semantics of N_3, a limited scope. It is therefore difficult to use them to reference the same class in different rules. We will give a more elaborate explanation in the next section. After that we will describe the translation step in more detail.

5.1 Grounding the Ontology

Before translating the TBox into rules we have to replace all blank nodes by URIs or literals. To understand the reason for this skolemization step, consider the example in Listing 2. The example contains triples which further describe

```
1  @prefix : <http://ontology/Accio.owl#>.
2  @prefix owl: <http://www.w3.org/2002/07/owl#>.
3  @prefix rdf: <http://www.w3.org/1999/02/22-rdf-syntax-ns#>.
4  @prefix rdfs: <http://www.w3.org/2000/01/rdf-schema#>.
5
6  :Call   rdfs:subClassOf [
7                           rdf:type owl:Class ;
8                           owl:intersectionOf (
9                                               :PatientTask
10                                              :UnplannedTask
11                                             )
12                         ] .
```

Listing 2. ACCIO example: a call is both, a patient task and an unplanned task.

the class :Call from Formula 2. A call is a patient task and an unplanned task, or to be more specific: the class :Call is subclass of an anonymous class which is the intersection of the classes :PatientTask and :UnplannedTask. Even though N_3 supports rules which contain blank nodes, it is exactly this anonymous class which causes problems. Being unlabeled, the blank node can be referred by an arbitrary new blank node name. A translation as done in Formula 4 would result in a rule like:

$$\{?X \text{ a } :Call.\} \Rightarrow \{?X \text{ a } \_:newblank.\}. \tag{5}$$

This rule means, that every instance of the class :Call is also instance of *some* other class. This knowledge can already be gained by Formula 4 and does not have much influence on further reasoning. And even if the blank node in Listing 2 would be labeled by, for example, _:intersection1 a new rule

$$\{?X \text{ a } :Call.\} \Rightarrow \{?X \text{ a } \_:intersection1.\}. \tag{6}$$

would have no other meaning than Formula 4 as in N_3 the scope of a blank node is always only the graph, i.e. the curly brackets { }, in which it occurs [3,5]. The consequence of the rule would not refer to our intersection of patient tasks and unplanned tasks.

We perform the grounding step by using the EYE reasoner. The reasoner provides the option to obtain a skolemized version of any input N_3 file(s). The switch --no-qvars replaces every blank node by a unique skolem IRI following the naming convention as described in the RDF specification [10]. It additionally makes sure that equally named blank nodes only get assigned the same skolem IRI if they actually refer to the same thing. Producing a grounded version of the ontology enables us in further reasoning steps to use the new identifiers for (formally) anonymous classes in different rules.

5.2 Translation Step

As explained above, the next step after having produced a grounded version of the ontology's TBox is to produce the new specialized rules. Here, we make use

```
1  @prefix rdfs: <http://www.w3.org/2000/01/rdf-schema#>.
2  @prefix rdf: <http://www.w3.org/1999/02/22-rdf-syntax-ns#>.
3
4  {?C rdfs:subClassOf ?D.} => {{?X a ?C.} => {?X a ?D.}.}.
```

Listing 3. Rule producing new rule for every occurrence of `rdfs:subClassOf`; based on the `rdfs:subClassOf` class axiom of Listing 1.

of a property of Notation3: rules can not only be applied to derive new triples but also to derive new rules. To illustrate that we consider a simple rule:

$$\{ \underbrace{\texttt{:Call rdfs:subClassOf :Task.}}_{\text{satisfied ontology triple(s)}} \} => \{ \underbrace{\{\texttt{?X a :Call}\}=>\{\texttt{?X a :Task.}\}.}_{\text{produced new rule(s)}} \}.$$

Just as simple rules enable the reasoner to derive new triples from the fact that its antecedent is fulfilled, the rule above, applied on Formula 1, derives a new rule, namely Formula 4. Nevertheless, the rule as stated above is too specific to be used for our purpose: if we already knew that the ontology contained the triple in Formula 1 we could also write the rule in Formula 4 directly instead of writing a rule which will surely produce it. Our rule needs to be more general as we want to handle all `owl:subclassOf` triples in that same way and always produce a rule similar to the rule expressed in Formula 4. This more general rule can be found in Listing 3. Applied on Formula 1 the variable `?C` gets unified with the URI `:Call` and the variable `?D` gets unified with `:Task`, thus, Rule 4 can be derived. Similarly, an application of the rule in Listing 3 on triple

> :UnplannedTask rdfs:subClassOf :Task.

results in a new rule

> {?X a :UnplannedTask.} => {?X a :Task.}.

The same principle can be applied for other OWL concepts. Listing 4 shows a rule[2] which handles the concept `owl:intersectionOf`. Note that this rule uses a built-in predicate of Notation3, `list:in`. A triple using `list:in` as a predicate is true if and only if the object is a list and the subject is an entry of that list. If we apply this rule to the (now skolemized) intersection expressed in Listing 2

> :InterClass1 owl:intersectionOf (:PatientTask :UnplannedTask).

two rules will be produced by that:

> {?x a :InterClass1} => {?x a :PatientTask.}.

and

---

[2] The rule is motivated by the cls-int2 rule in Table 6 on [9].

```
1  @prefix list: <http://www.w3.org/2000/10/swap/list#>.
2  @prefix owl: <http://www.w3.org/2002/07/owl#>.
3
4  {?C owl:intersectionOf ?L. ?D list:in ?L} =>
5                                {{?X a ?C.} => {?X a ?D}}.
```

Listing 4. Rule-producing rule for `owl:intersectionOf`.

$$\{?x \ a \ :InterClass1\} \ => \ \{?x \ a \ :UnplannedTask.\}.$$

The above example illustrates another useful property of Notation3: Notation3 treats lists themselves, not only their reified version, as elements of the language. There are many built-in predicates which enable the user to write clear rules regarding lists and to refer to all elements of a given list. For working with OWL ontologies this is a real advantage as lists are normally used together with many OWL concepts like the above `owl:intersectionOf` or for example `owl:unionOf`.

To produce new rules by applying the rules described above, the rule producing rules have to be applied as filter rules for the reasoner. Notation3 reasoners normally take one ore more input files—consisting of rules and facts—and a query file containing rules into account. Based in the input files the reasoner outputs the logical consequences of the filter rules. In our present case these are the specialized rules. The rules produced by the two described steps do now replace the TBox of the ontology and can be used for further reasoning.

6   Evaluation

The aforementioned methodology replaces generic and complex constructs in the TBox by specialized rules that provide the same functionality. To test how much performance we gain by using this pre-processing step we tested a scenario of our use case with two rule sets: the first traditional rule set [11] processes the triples of the original TBox while reasoning and acts on top of those together with the actual ABox data, the second precomputed rule set contains the specialized rules which already take all TBox triples into account, therefore in this case the original TBox is not needed for further reasoning. All experiments were run on the same technology stack[3].

6.1   Ontology and Data

To represent the data as described above we make use of the ACCIO ontology which was designed to represent all aspects of patient care in a hospital. The ontology contains ca. 3,500 triples (414 named classes, 157 object properties, 38 data type properties). A full description is given by Ongenae et al. [15].

---

[3] Hardware: Intel(R) Xeon(R) E5620@2.40 GHz CPU with 12 GB RAM. Software: Debian "Wheezy", EYE-Autumn15 09261046Z and SWI-Prolog 6.6.6.

This ontology was filled with data describing wards in a hospital. This data was simulated, based on real-life situations, as deducted from user studies [15]. The data was scaled by increasing the amount of wards from 1 to 10 to fill the ABox with more data. The description of such a ward contains approximately 1,000 static triples. Additionally, there was dynamic data such as for example the location of nurses or the status of calls taken into account.

6.2 Test Scenario

We compared the reasoning times of the two rule sets by running a scenario, based on a real-life situation. This scenario consists of a sequence of events, which we list below, where the expected outcome of the reasoning is indicated in brackets.

1. A patient launches a call (*assign nurse and update call status*)
2. The assigned nurse indicates that she is busy (*assign other nurse*)
3. The newly assigned nurse accepts the call task (*update call status*)
4. The nurse moves to the corridor (*update location*)
5. The nurse arrives at the patients' room (*update location, turn on lights and update nurse status*)
6. The nurse logs into the room's terminal (*update status call and nurse, open lockers*)
7. The nurse logs out again (*update status call and nurse, close lockers*)
8. The nurse leaves the room (*update location and nurse status and turn off lights*)

6.3 Results

The aforementioned scenario was run 35 times, consisting of 3 warm-up runs and 2 cool-down runs, for 1 ward and 10 wards, for both rule sets. By averaging the 30 remaining reasoning times per amount of wards and per rule set, we provide the results as shown as a table in Fig. 1, and depicted in Fig. 2.

The figures show how preprocessing the rules improves reasoning times significantly, consistently requiring only a quarter of the reasoning time. This trend manifests itself regardless of the amount of dynamic data involved. Whereas the traditional rule set can no longer be used in a hospital with 10 wards the preprocessed rule set still provides reasonable reasoning times.

wards	1 ward								10 wards							
event	1	2	3	4	5	6	7	8	1	2	3	4	5	6	7	8
traditional	2.1	2.1	2.4	2.1	2.1	2.2	2.4	2.1	30.7	30.7	34.9	30.6	30.6	30.7	35.0	30.5
preprocessed	0.4	0.6	0.7	0.4	0.4	0.4	0.5	0.4	6.8	10.7	12.2	8.1	8.0	6.7	9.3	8.1

Fig. 1. Reasoning times using traditional rules and preprocessed rules in seconds. Preprocessing significantly reduces reasoning times.

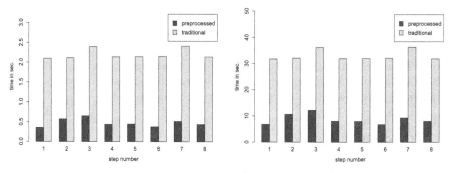

(a) 1 ward, reasoning time per event.   (b) 10 wards, reasoning time per event.

Fig. 2. Comparison of reasoning times using preprocessed and traditional rules. The preprocessing step improves reasoning times.

7   Conclusion and Future Work

In this paper we have shown that precomputing and using specialized rules based on the ontology's TBox improve reasoning times of OWL RL reasoning by about 75 %. The main cause for that is that the newly computed rules are less complex—in terms of the variables which have to be unified during reasoning—than the original version of the rules taken from the OWL RL profile description. Another aspect which makes reasoning faster in our set up is the fact that rules for concepts which are not even present in the ontology's TBox will not get produced by the preprocessing step. If for example the rather expensive concept owl:sameAs does not occur in any triple of the considered ontology, no specialized rules will be produced for this concept.

The presented preprocessing step consists of two simple reasoning runs which can be performed before dealing with additional input data. Using the EYE reasoner this preprocessing normally takes only a few seconds. In set ups where the TBox does not change during run time the produced rules can be used whenever the ABox data to reason on changes as in the example introduced in this paper. Our approach is independent of the reasoning done on top of the TBox by additional rules. This makes the rule version of the ontology's TBox even more suitable for reuse.

Our approach makes use of the special properties of Notation3 Logic. By providing the option of using rules to produce new rules this logic is particularly suitable for our purposes. Furthermore Notation3 offers multiple predicates to act on lists as for example the function list:in. This eases the implementation of rule producing rules based on OWL predicates as there are many OWL constructs which are normally stated with lists in their object position.

Notation3 Logic posses other interesting properties which we are planning to apply in future work: in N_3, rules can have existential variables in their consequence. Using this particular property it will be possible to also cover OWL EL concepts which are not present in OWL RL. Similarly as done in for example

OWLim [7] we are planning to include OWL EL in our implementation. We furthermore want to investigate the actual costs of processing the different OWL concepts by our newly produced rules. This will enable us to recommend the exclusion of particular concepts if not really needed.

Acknowledgements. The research activities described in this paper were funded by Ghent University, iMinds, the IWT Flanders, the FWO-Flanders, and the European Union, in the context of the project "ORCA", which is a collaboration of Televic Healthcare, Internet-Based Communication Networks and Services (IBCN), and Data Science Lab (DSLab).

References

1. FuXi 1.4: A Python-based, bi-directional logical reasoning system for the semantic web. http://code.google.com/p/fuxi/
2. Arndt, D., et al.: Ontology reasoning using rules in an eHealth context. In: Bassiliades, N., Gottlob, G., Sadri, F., Paschke, A., Roman, D. (eds.) RuleML 2015. LNCS, vol. 9202, pp. 465–472. Springer, Heidelberg (2015)
3. Arndt, D., Verborgh, R., De Roo, J., Sun, H., Mannens, E., Van De Walle, R.: Semantics of notation3 logic: a solution for implicit quantification. In: Bassiliades, N., Gottlob, G., Sadri, F., Paschke, A., Roman, D. (eds.) RuleML 2015. LNCS, vol. 9202, pp. 127–143. Springer, Heidelberg (2015)
4. Berners-Lee, T.: *cwm* (2000–2009). http://www.w3.org/2000/10/swap/doc/cwm.html
5. Berners-Lee, T., Connolly, D.: Notation3 (N3): A readable RDF syntax. In: W3C Team Submission, March 2011. http://www.w3.org/TeamSubmission/n3/
6. Berners-Lee, T., Connolly, D., Kagal, L., Scharf, Y., Hendler, J.: N3 Logic: a logical framework for the World Wide Web. Theory Pract. Logic Program. **8**(3), 249–269 (2008)
7. Bishop, B., Bojanov, S.: Implementing OWL 2 RL and OWL 2 QL rule-sets for OWLIM. In: OWLED, vol. 796 (2011)
8. Bishop, B., Kiryakov, A., Ognyanoff, D., Peikov, I., Tashev, Z., Velkov, R.: OWLIM: a family of scalable semantic repositories. Semant. Web **2**(1), 33–42 (2011)
9. Calvanese, D., Carroll, J., Di Giacomo, G., Hendler, J., Herman, I., Parsia, B., Patel-Schneider, P.F., Ruttenberg, A., Sattler, U., Schneider, M.: OWL 2 web ontology language profiles 2nd edn. In: W3C Recommendation, December 2012. www.w3.org/TR/owl2-profiles/
10. Cyganiak, R., Wood, D., Lanthaler, M.: RDF 1.1: Concepts and Abstract Syntax. In: W3C Recommendation, February 2014. http://www.w3.org/TR/2014/REC-rdf11-concepts-20140225/
11. De Roo, J.: Euler yet another proof engine (1999–2015). http://eulersharp.sourceforge.net/
12. De Roo, J.: EYE and OWL 2 (1999–2015). http://eulersharp.sourceforge.net/2003/03swap/eye-owl2.html
13. Declerck, T., Krieger, H.U.: Translating XBRL into description logic. an approach using protege, sesame & OWL. In: BIS, pp. 455–467 (2006)

14. Horrocks, I., Patel-Schneider, P.F., Boley, H., Tabet, S., Grosof, B., Dean, M.: SWRL: a semantic web rule language combining OWL and RuleML. In: W3C Member Submission, 21 May 2004. http://www.w3.org/Submission/SWRL/

15. Ongenae, F., Bleumes, L., Sulmon, N., Verstraete, M., Van Gils, M., Jacobs, A., De Zutter, S., Verhoeve, P., Ackaert, A., De Turck, F.: Participatory design of a continuous care ontology: towards a user-driven ontology engineering methodology. In: Proceedings of the Knowledge Engineering and Ontology, pp. 81–90 (2011)

16. Parsia, B., Sirin, E.: Pellet: An OWL DL reasoner. In: Proceedings of the Third International Semantic Web Conference (2004)

17. Patel, C., et al.: Matching patient records to clinical trials using ontologies. In: Aberer, K., et al. (eds.) ASWC 2007 and ISWC 2007. LNCS, vol. 4825, pp. 816–829. Springer, Heidelberg (2007)

18. Shearer, R., Motik, B., Horrocks, I.: Hermit: a highly-efficient OWL reasoner. In: OWLED, vol. 432, p. 91 (2008)

19. Verborgh, R., De Roo, J.: Drawing conclusions from linked data on the web. IEEE Softw. **32**(5), 23–27 (2015)

20. Wu, Z., Eadon, G., Das, S., Chong, E.I., Kolovski, V., Annamalai, M., Srinivasan, J.: Implementing an inference engine for RDFS/OWL constructs and user-defined rules in Oracle. In: 2008 IEEE 24th International Conference on Data Engineering, ICDE 2008, pp. 1239–1248. IEEE (2008)

On the Capabilities and Limitations of OWL Regarding Typecasting and Ontology Design Pattern Views

Adila A. Krisnadhi[1,3]([✉]), Pascal Hitzler[1], and Krzysztof Janowicz[2]

[1] Wright State University, Dayton, OH, USA
Krisnadhi.2@wright.edu
[2] University of California, Santa Barbara, CA, USA
[3] Universitas Indonesia, Depok, Indonesia

Abstract. In ontology engineering, particularly when dealing with heterogeneous domains and their subfields, legacy data, various data models, existing standards, code lists, and so forth, there is a frequently recurring need to express certain types of axioms that allow diverse representational choices interoperate. Some of these axioms, which we call typecasting axioms, point to limitations of the Web Ontology Language (OWL), while others require best practice guides for the community. Here, we introduce these typecasting axioms and elaborate how such axioms can help the development of data integration using ontology and ontology patterns. We then conclude with a brief catalog of open research problems motivated by typecasting axioms, which may be of potential interest to both application developers and researchers working on logical foundations of OWL.

1 Introduction and Motivation

During our ontology engineering work with subject matter experts from a wide range of domains including the broader geo-sciences [10], industrial ecology, the digital humanities, libraries and the publishing industry, particle physics, and so forth, we became aware of the recurring need to express certain types of axioms necessary to bridge diverse representational choices, and thus enable interoperability between them in the same ontological framework. Often, these axioms can be easily expressed using first-order predicate logic, but the description logic (DL) underlying the Web Ontology Language OWL [7] does not – or not obviously – enable us to express these axioms.

Our goal is twofold: first, to motivate and describe these types of axioms as well as the capabilities and limitations of OWL to represent them, to the extent we are aware of them; and second, to highlight these limitations as open problems on which researchers interested in improving and extending OWL and its underlying logics could work. In this paper, we particularly focus on axioms called *typecasting* axioms, which allow one to seamlessly switch between class-centric, individual-centric, and property-centric representation. These axioms are not only relevant to ontology modeling, but also in ontology alignment with

© Springer International Publishing Switzerland 2016
V. Tamma et al. (Eds.): OWLED 2015, LNCS 9557, pp. 105–116, 2016.
DOI: 10.1007/978-3-319-33245-1_11

complex mappings. Some of the issues herein were already alluded to by Noy [16], however they were not discussed in the context of formal semantics of OWL and the underlying description logics-based formalisms.

Our discussion is structured as follows. Section 2 introduces typecasting axioms and discusses how and why OWL can or cannot represent them. This is followed by Sect. 3 where we further discuss a particular use of typecasting axioms in the context of Ontology Design Patterns [6] to express *view* expansion and contraction, which are useful in the context of linked data publishing and integration [10,12,17]. We then conclude in Sect. 4 with a list of research questions compiled from throughout the paper.

2   Typecasting in OWL

In this section we discuss three kinds of typecasting that we frequently encountered in our work on ontological modeling. Casting between types is the implicit or explicit process by with one (data) type is converted into another type, e.g., widening an *int* to a *long*. In object-oriented modeling, this includes accessing objects that instantiate certain types as objects of their common ancestor type. To give a simple example, a Point Of Interest (POI) class may define a method to return the spatial footprint of the place as geographic (point) coordinates. All classes that extend the POI class, say Restaurant and Hospital, can be queried for their footprint by iterating over a collection of POI.

Here, we use the term typecasting to refer to translation between multiple representational choices to define a notion using either individuals, classes, or properties in the context of description logics. This is part of a bigger picture that we call *ontology virtualization* by which an underlying model can be exposed in different ways to suit particular needs or paradigms.

2.1   Typecasting Individual to Class and Back: Explicit Versus Implicit Typing of Instances

The first case is concerned with the representational choice between the explicit typing of individuals (via rdf:type) versus the identification of the type of an individual by reference to a classname, given as an individual. In other words, we do typecasting from a class to an individual and vice versa.

Problem Description. Schematically, the two representational choices are depicted in Fig. 1. Case 1a at the top part of the figure corresponds to the explicit use of `rdf:type` to assert a type of an individual, which, generally speaking, seems to be more in the spirit of OWL. Case 1b at the bottom part of the figure, however, sometimes seem more natural for domain experts, e.g., when referring to an externally controlled vocabulary.

Consider, for example, the case of measurement types. A concrete measurement of a particular characteristic of a feature of interest, e.g., a lake, can be

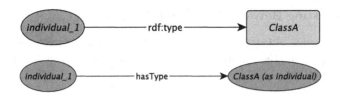

Fig. 1. Axiom (1) maps Case 1a (top part) to Case 1b (bottom part), while axiom (2) maps in the opposite direction. Blue nodes are OWL named individuals, while the yellow node is an OWL class (Color figure online).

of type NitrateConcentration (which in turn is a Concentration measurement). This can be asserted through the following triple.

<code>ex:measurement1 rdf:type geo:NitrateConcentration.</code>

We assume here that the namespace `geo:` refers to an appropriate ontology that contains measurement types. At the same time, however, it may be appropriate for this ontology to incorporate an existing controlled vocabulary for the identification of measurement types widely used in certain fields of the geosciences. Such controlled vocabularies often come in the form of code lists or may describe measurement types as individuals because one may want to give additional information about them, say

<code>geo:NO3-concentration geo:potentialSource geo:urban_runoff.</code>

With `geo:NO3-concentration` being an individual, one would therefore appropriately identify the measurement type for `:measurement1` by specifying

<code>ex:measurement1 geo:hasMeasurementType geo:NO3-concentration.</code>

Logical Aspects. This typecasting case can be handled easily in OWL. Axiom (1) maps Case 1a to 1b, while axiom (2) maps Case 1b to 1a.

$$\text{ClassName} \sqsubseteq \exists \text{hasType}.\{\text{classname}\} \tag{1}$$

$$\exists \text{hasType}.\{\text{classname}\} \sqsubseteq \text{ClassName} \tag{2}$$

For the nitrate concentration example, the mappings in both direction can be expressed by:

<code>geo:NitrateConcentration ≡</code>
<code>∃geo:hasMeasurementType.{geo:NO3-concentration}.</code>

It shall be noted that the above case is closely related to punning between classes and individuals, i.e., the use of one identifier to denote both a class name and an individual name, which is allowed in OWL 2 DL. In fact, in the above example, we could have used `geo:NO3-concentration` as a class name in addition to using it as an individual name (or only using `geo:NitrateConcentration` for that matter). Typecasting is, however, more general in the sense that it allows one to use different identifier to refer to the same typing of an instance and employ any object property as a typing predicate, simulating `rdf:type`.

2.2   Typecasting Between Class and Property

The next two kinds of typecasting is concerned with the representational choice between using a simple property or using a class to represent a relationship between two entities.

Problem Description. Schematically, the two representational choices are depicted in Fig. 2. Here, Case 2a simply uses a property to represent a relationship between two individuals, while Case 2b uses a reified representation of the relation, which is actually a (possibly non-atomic) class.

For example, consider the set of triples in Fig. 3 stating that an oceanographic cruise has a lead scientist provided as an individual `ex:PeterWiebe`. These triples in Turtle syntax corresponds to Case 2a (disregarding the instance typing triples).

Meanwhile, the set of triples in Fig. 4 corresponds to Case 2b where essentially the same relationship as above is represented using a reified representation.

The representational choice as described by Case 2b is in fact very common, e.g. as part of the so-called Agent Role ontology design pattern,[1] and has in some form even be adopted by schema.org under the term *role*.[2] The advantage of the second, more verbose representation is that additional information can be added to the blank node, e.g., the funding agency, affiliation, and so forth. Note that

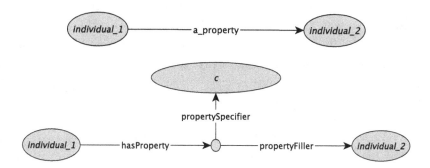

Fig. 2. Case 2a (top) and Case 2b (bottom). In the bottom part, the unlabeled blue node can either be a named or anonymous individual (blank node) (Color figure online).

```
ex:cruise123    geo:hasLeadScientist    ex:PeterWiebe ;
                rdf:type                geo:Cruise .
ex:PeterWiebe   rdf:type                geo:Person .
```

Fig. 3. Triples for Case 2a

---

[1] http://ontologydesignpatterns.org/wiki/Submissions:AgentRole.
[2] https://schema.org/Role.

```
ex:cruise123  rdf:type    geo:Cruise .
ex:cruise123  geo:hasParticipation  [
                  rdf:type geo:LeadScientistRole ;
                  geo:playsRole   geo:LeadScientist ;
                  geo:isPlayedBy  ex:PeterWiebe ] .
ex:PeterWiebe rdf:type    geo:Person .
```

Fig. 4. Triples for Case 2b

the typecasting discussed in Sect. 2.1 can also be applied to the individual acting as the property specifier in case 2b above.

Logical Aspects: Rolification. Mapping from Case 2b to 2a corresponds to typecasting from class to property. Note that Case 2b is indeed a class-centric representation because the intended relationship is represented through the unlabeled node in the middle of the bottom part of Fig. 2, which is an instance of the non-atomic class expression $\exists$propertySpecifier.$\{c\}$. Obviously, one can assign a class name to such an expression if so desired. Also, `rdf:type` and a class name can be used instead of propertySpecifier and c.

The typecasting from class to property desired above employs a technique called *rolification* [11,18].[3] This is a key technique for representation of (Datalog) rules in OWL. There are of course rules that can be readily expressed using OWL axioms, e.g., guarded domain restrictions such as

$$\text{Person}(y) \wedge \text{hasLeadScientist}(x, y) \rightarrow \text{Cruise}(x),$$

which is equivalent to

$$\exists\text{hasLeadScientist.Person} \sqsubseteq \text{Cruise}$$

However, this is not the case for other rules such as:

$$\text{Cruise}(x) \wedge \text{hasParticipation}(x, y) \wedge \text{LeadScientistRole}(y) \wedge \text{isPlayedBy}(y, z)$$
$$\wedge \text{ Person}(z) \rightarrow \text{hasLeadScientist}(x, z)$$

The above rule is in fact what we need to map Case 2b to 2a. It can be expressed in OWL by *rolifying* the three class names, i.e. by introducing new properties R_{Cruise}, $R_{\text{LeadScientistRole}}$, and R_{Person} and asserting the following axioms:

$$\text{Cruise} \equiv \exists R_{\text{Cruise}}.\text{Self}, \qquad \text{Person} \equiv \exists R_{\text{Person}}.\text{Self}$$
$$\text{LeadScientistRole} \equiv \exists R_{\text{LeadScientistRole}}.\text{Self}$$
$$R_{\text{Cruise}} \circ \text{hasParticipation} \circ R_{\text{LeadScientistRole}} \circ \text{isPlayedBy} \circ R_{\text{Person}} \sqsubseteq \text{hasLeadScientist}$$

Notice that the three class names were typecasted into property names through rolification axioms of the form $A \equiv \exists R_A.\text{Self}$.

---

[3] In DL literature, properties are called roles, hence the term rolification; not to be confused with role as defined by the Agent Role pattern or schema.org.

More generally, mapping Case 2b to 2a in Fig. 2 is expressed using the following rule

$$\text{hasProperty}(x, y) \wedge \text{propertySpecifier}(y, c) \wedge \text{propertyFiller}(y, z)$$
$$\rightarrow \text{a\_property}(x, z),$$

which can be expressed in OWL using the two axioms

$$\exists \text{propertySpecifier}.\{c\} \sqsubseteq \exists \text{propertySpecifier\_c.Self}$$
$$\text{hasProperty} \circ \text{propertySpecifier\_a} \circ \text{propertyFiller} \sqsubseteq \text{a\_property}$$

where the complex class $\exists \text{propertySpecifier}.\{c\}$ is typecasted into the property propertySpecifier_c via a rolification axiom.

Since rolification allows one to typecast class into property, an additional benefit of the use of rolification is that it allows us to express *typed property chains* of the form

$$R_1 \circ \ldots \circ R_n \sqsubseteq R,$$

where each R_i is either a property name or a class name, the latter of which is to be typecasted into a property using rolification.

There is however a caveat in using rolification axiom. While it is of course expressible in OWL DL, its primary use cases, i.e., general conversion of rules, always involve property chains. For them, OWL DL imposes a so-called *regularity restriction* on the use of property chains [8], which may be violated by the introduced ones. The origin of the regularity restriction is that without it, reasoning over the logic would be undecidable. While this means that the restriction cannot be lifted in its entirety without rendering the logic undesirable, it would be helpful to soften it, i.e., to describe types of cases which violate regularity, but which retain decidability. In addition, approximate work-arounds are possible, e.g., using so-called *nominal schemas* [11,13].

Logical Aspects: Reification. Mapping in the other direction, i.e., from case 2a to case 2b, corresponds to typecasting property to class. For example, given a set of triples in Fig. 3, we want to express an axiom that allows us to infer triples in Fig. 4. This amounts to a well-known modeling technique called *reification*. The axiom cannot be expressed in OWL, but can be handled using rules with existential head – well-known in database as tuple-generating dependencies (TGDs) [2]. In the context of Fig. 2, the TGD is of the following form:

$$\text{a\_property}(x, z) \rightarrow \exists y.(\text{hasProperty}(x, y) \wedge \text{propertySpecifier}(y, a)$$
$$\wedge \text{propertyFiller}(y, z))$$

Note that using rolification axioms is not sufficient because they cannot allow us to infer the existence of the new node for the reification (the RDF blank node in Fig. 4).

3   Ontology Design Pattern View Contraction and Expansion

When developing ontologies or ontology design patterns (ODPs) for the purpose of data integration, ontology engineers often have to introduce complex structures like reified relationships to cover the richness of the data being integrated or to provide flexibility in the integrating schema. However, from the perspective of a particular user or data provider, such complications may not be desirable. For them, simplified version of the global schema, which can be specially tailored to be sufficient for their needs may be preferable. In the context of ontology-based or ODP-based data integration, such a simplified version of the global schema corresponds to what we call a *view* by which we mean a set of shortcuts through an ontology or an ontology design pattern. To illustrate the concept, which is discussed also in [12,17], we adapt an example from the GeoLink oceanography ontology [10].

Referring to Fig. 5, the red arrows indicate shortcuts, and we will discuss the case of the isTraversedBy shortcut. Of course the picture is only a visualization of a part of the ontology, which consists of a set of OWL axioms which we do not list here.

In the fabric of the ontology, the isTraversedBy shortcut is in fact redundant, i.e. it can be inferred using the rule

$$\mathrm{Vessel}(x) \wedge \mathrm{isUndertakenBy}(y,x) \wedge \mathrm{Cruise}(y)$$
$$\wedge \mathrm{hasTrajectory}(y,z) \wedge \mathrm{Trajectory}(z) \wedge \mathrm{hasSegment}(z,w)$$
$$\wedge \mathrm{Segment}(w) \rightarrow \mathrm{isTraversedBy}(w,x). \quad (3)$$

Since the application of the rule results in a simpler representation of the relationship between a trajectory segment and the vessel traversing it, we refer to this type of rule also as a *contraction*.

The reverse of a contraction is an *expansion*. In our experience, this case occurs when, e.g., a data provider may have only information about trajectories (and their segments) which oceanographic vessels have taken. In order to populate the ontology with this data, it is required to *expand* the data by inserting an additional individual (or a blank node) as the cruise connecting the trajectory and the vessel.

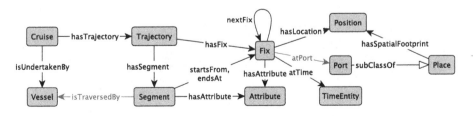

Fig. 5. Part of the GeoLink oceanography ontology to illustrate views.

3.1 Contraction

A generic depiction of the view idea is presented in Fig. 6. The grey ellipse shall indicate a labeled graph which in turn can be represented as a conjunction of unary and binary predicates involving ClassA and ClassB, like

$$\text{ClassA}(x) \wedge \text{ClassB}(y) \wedge C_1(x_1) \wedge \cdots \wedge C_n(x_n) \wedge R_1(y_1, y_2) \wedge \cdots \wedge R_{l_0}(y_{l_0}, y_{l_0+1}),$$

where x, y, the x_i and the y_j are any variables.

Contraction (i.e., a shortcut between the classes ClassA and ClassB) can then be expressed using the rule

$$\text{ClassA}(x) \wedge \text{ClassB}(y) \wedge C_1(x_1) \wedge \cdots \wedge C_n(x_n) \wedge R_1(y_1, y_2) \wedge \cdots \wedge R_k(y_k, y_{k+1})$$
$$\rightarrow \text{shortcut}(x,y). \tag{4}$$

Note that the simpler typecasting case discussed in Sect. 2.2 can in fact be understood as a very simple case of contraction and expansion.

The rule expressing a shortcut (i.e., contraction) cannot in general be represented in OWL, and this is well-known. In particular, if the graph representing the rule body is cyclic, this is not possible in many cases. Discussing this in detail is out of scope for this paper, but a detailed account of this can e.g. be found in [11,13].

However, let us work with the earlier example from rule (3), which is not cyclic. In this case we can convert the rule into OWL using rolification, which results in the following set of axioms.

$$\text{Vessel} \equiv \exists R_{\text{Vessel}}.\text{Self}, \qquad \text{Cruise} \equiv \exists R_{\text{Cruise}}.\text{Self}$$
$$\text{Trajectory} \equiv \exists R_{\text{Trajectory}}.\text{Self}, \qquad \text{Segment} \equiv \exists R_{\text{Segment}}.\text{Self}$$
$$R_{\text{Segment}} \circ \text{hasSegment}^- \circ R_{\text{Trajectory}} \circ$$
$$\circ \text{hasTrajectory}^- \circ R_{\text{Cruise}} \circ \text{isUndertakenBy} \sqsubseteq \text{isTraversedBy}$$

The problem is again, of course, that the introduction of additional role chains may render the ontology to be outside OWL DL due to possible violations of regularity restrictions.

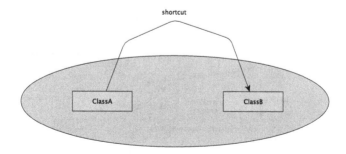

Fig. 6. Depiction of a generic shortcut.

The cases where rules are cyclic pose particular challenges. We illustrate this by an example taken from [11]. The rule defines a shortcut indicating a review assignment with a conflict of interest.

$$\text{hasReviewAssignment}(v, x) \wedge \text{hasAuthor}(x, y) \wedge \text{atVenue} \wedge$$
$$\wedge\, \text{hasSubmittedPaper}(v, u) \wedge \text{hasAuthor}(u, y) \wedge \text{atVenue}(u, z)$$
$$\rightarrow \text{hasConflictingAssignedPaper}(v, x)$$

Approximate (sound but incomplete) translations of such cyclic rules into OWL are possible using DL-safe rules [9,15]. Better approximations (i.e., with not as much loss in terms of logical consequences) are possible using so-called *nominal schemas* [11,13]. While in the meantime some results have been obtained regarding efficient reasoning with nominal schemas [5,19], the topic does still require in depth exploration to obtain sufficient coverage for modeling purposes.

Assuming familiarity with rule to OWL conversion techniques as discussed e.g. in [11], we identify several research questions which address such conversion issues. Some of them have in fact already been exposed by our earlier discussions.

(1) Translation of rules usually requires rolification and the use of role chains, i.e. softening regularity restrictions would be extremely helpful.

(2) Approximate translation of rules (approximate in order to avoid regularity issues) currently requires the use of nominal schemas, for which efficient reasoning algorithms, as well as suitable modeling and reasoning tools, require further investigation.

(3) Translated rules often fall into the OWL EL fragment with additional use of inverse roles. While OWL EL requires regularity, the regularity requirement is not required for decidability of the logic. However, in the presence of inverse roles, together with a non-regular set of role chains, the logic becomes undecidable [1]. Softening the regularity requirement for OWL EL with addtional inverse roles would make it possible to translate more shortcut rules.

(4) Likewise, OWL EL (with the regularity restriction) together with inverse roles is no longer tractable. Research into conditions under which tractability is retained would be helpful in practice – see e.g. for [4] for some work related to this issue.

Another issue arising out of shortcuts is if ClassB, in our generic example, is actually a datatype, i.e., the infered role 'shortcut' shall be a datatype property. Structurally, representation of the corresponding rule should follow the same method, however the resulting OWL axioms will then usually involve a role chain with a datatype property as the final, right-most role. However, the OWL standard currently does not allow this. We conjecture that allowing this would probably be a minor extension of the standard, but this still requires looking at.

3.2   Expansion

Expansion is the reverse of contraction, i.e. expanding from a shortcut into the graph, as in our generic example. It can be understood as a generalized version of the direction from Case 2a to 2b in Sect. 2.2 where a blank node is introduced, i.e. as a type of role introduction instead of using an elementary property.

Of course, simply reversing the implication arrow in rule (4) is insufficient, as quantification of the variables needs to be addressed. The appropriate axiomatization, in fact, is the following.

$$\text{shortcut}(x,y) \rightarrow \text{ClassA}(x) \land \text{ClassB}(y) \land \exists x_1 \ldots \exists x_n \exists y_1 \ldots \exists y_n (C_1(x_1) \land \ldots$$
$$\cdots \land C_n(x_n) \land R_1(y_1, y_2) \land \cdots \land R_k(y_k, y_{k+1}))$$

Similar to the case above in Sect. 2.2, existential rules appear to be a suitable paradigm, in principle. However the potentially rather complex rule heads deserve considerable investigation, in particular if it is to be integrated with ontology reasoning.

A specific case which may also deserve studying is when the rule head may be translatable into a right-hand-side role chain, i.e. an axiom of the form

$$R \sqsubseteq R_1 \circ \cdots \circ R_n,$$

possible after some rolification. Right-hand-side role chains have been studied in the literature and in the general case they lead to undecidability, particularly when left-hand side role chains are present. Decidability by generalizing regularity restriction where shown by Mosurovic, et al. [14]. On the other hand, the rule above can also be categorized into guarded TGDs [3] for which query answering is decidable. Note, however, that adding existential rules to OWL in general may cause the violation of guardedness condition, hence may not guarantee decidability.

4   Conclusions

We have seen that modeling issues arising in practice give rise to logical axioms which are currently not expressible within the OWL DL standard, and this prompts research questions which may ultimately lead to a suitable coverage in a later version of the standard. To provide an overview, we list the research questions raised by our discussion.

1. Relaxing RBox regularity constraints to make use of rolification easier, for several of the aspects mentioned above.
2. Relaxing RBox regularity constraints in the specific case of OWL EL with additional inverse roles would allow for the expression of more view contractions. Aspects to be considered would be both, decidability and tractability.
3. Develop more efficient reasoning algorithms and implementations for nominal schemas, as they are one way to circumvent the regularity issues arising from rolification.

4. Investigate reasoning aspects regarding role chains which end in datatype literals, including the issue of right-hand-side role chains.
5. Investigate right-hand-side role chains as a possible extension of OWL DL.
6. Investigate the integration of existential rules with OWL DL, in particular for complex rule heads.

Acknowledgements. This work was supported by the National Science Foundation under award 1017225 *III: Small: TROn – Tractable Reasoning with Ontologies* and award 1440202 *EarthCube Building Blocks: Collaborative Proposal: GeoLink – Leveraging Semantics and Linked Data for Data Sharing and Discovery in the Geosciences.*

References

1. Baader, F., Brandt, S., Lutz, C.: Pushing the EL envelope. In: Kaelbling, L.P., Saffiotti, A. (eds.) IJCAI-2005, Proceedings of the Nineteenth International Joint Conference on Artificial Intelligence, Edinburgh, Scotland, UK, 30 July–5 August, 2005, pp. 364–369 (2005)
2. Beeri, C., Vardi, M.Y.: A proof procedure for data dependencies. J. ACM **31**(4), 718–741 (1984)
3. Calì, A., Gottlob, G., Kifer, M.: Taming the infinite chase: query answering under expressive relational constraints. J. Artif. Intell. Res. (JAIR) **48**, 115–174 (2013)
4. Carral, D., Feier, C., Grau, B.C., Hitzler, P., Horrocks, I.: *EL*-ifying ontologies. In: Demri, S., Kapur, D., Weidenbach, C. (eds.) IJCAR 2014. LNCS, vol. 8562, pp. 464–479. Springer, Heidelberg (2014)
5. Carral, D., Wang, C., Hitzler, P.: Towards an efficient algorithm to reason over description logics extended with nominal schemas. In: Faber, W., Lembo, D. (eds.) RR 2013. LNCS, vol. 7994, pp. 65–79. Springer, Heidelberg (2013)
6. Gangemi, A.: Ontology design patterns for semantic web content. In: Gil, Y., Motta, E., Benjamins, V.R., Musen, M.A. (eds.) ISWC 2005. LNCS, vol. 3729, pp. 262–276. Springer, Heidelberg (2005)
7. Hitzler, P., Krötzsch, M., Parsia, B., Patel-Schneider, P.F., Rudolph, S. (eds.) OWL 2 Web Ontology Language: Primer. W3C Proposed Recommendation 22 September 2009 (2009). http://www.w3.org/TR/owl2-primer/
8. Hitzler, P., Krötzsch, M., Rudolph, S.: Foundations of Semantic Web Technologies. CRC Press, Chapman & Hall, Boca Raton (2010)
9. Hitzler, P., Parsia, B.: Ontologies and rules. In: Staab, S., Studer, R. (eds.) Handbook on Ontologies. International Handbooks on Information Systems, pp. 111–132. Springer, Heidelberg (2009)
10. Krisnadhi, A., et al.: The GeoLink modular oceanography ontology. In: Arenas, M., et al. (eds.) ISWC 2015. LNCS, vol. 9367, pp. 301–309. Springer, Heidelberg (2015). doi:10.1007/978-3-319-25010-6_19
11. Krisnadhi, A., Maier, F., Hitzler, P.: OWL and rules. In: Polleres, A., d'Amato, C., Arenas, M., Handschuh, S., Kroner, P., Ossowski, S., Patel-Schneider, P. (eds.) Reasoning Web 2011. LNCS, vol. 6848, pp. 382–415. Springer, Heidelberg (2011)

12. Krisnadhi, A.A., Hu, Y., Janowicz, K., Hitzler, P., Arko, R., Carbotte, S., Chandler, C., Cheatham, M., Fils, D., Finin, T., Ji, P., Jones, M., Karima, N., Lehnert, K., Mickle, A., Narock, T., O'Brien, M., Raymond, L., Shepherd, A., Schildhauer, M., Wiebe, P.: The GeoLink framework for pattern-based linked data integration. In: Proceedings of the ISWC 2015 Posters & Demonstrations Track a track within the 14th International Semantic Web Conference, ISWC 2015, Bethlehem, PA, USA, October 13, 2015 (2015)

13. Krötzsch, M., Maier, F., Krisnadhi, A., Hitzler, P.: A better uncle for OWL: nominal schemas for integrating rules and ontologies. In: Srinivasan, S., Ramamritham, K., Kumar, A., Ravindra, M.P., Bertino, E., Kumar, R. (eds.) Proceedings of the 20th International Conference on World Wide Web, WWW 2011, Hyderabad, India, March 28 – 1 April, 2011, pp. 645–654. ACM (2011)

14. Mosurovic, N., Krdzavac, H., Graves, M., Zakharyaschev, M.: A decidable extension of SROIQ with complex role chains and unions. J. Artif. Intell. Res. **47**, 809–851 (2013)

15. Motik, B., Sattler, U., Studer, R.: Query answering for OWL-DL with rules. J. Web Semant. **3**(1), 41–60 (2005)

16. Noy, N. (ed.) Representing classes as property values on the semantic web. W3C Working Group Note, 5 April 2005. http://www.w3.org/TR/swbp-classes-as-values/

17. Rodríguez-Doncel, V., Krisnadhi, A.A., Hitzler, P., Cheatham, M., Karima, N., Amini, R.: Pattern-based linked data publication: the linked chess dataset case. In: Hartig, O., Sequeda, J., Hogan, A. (eds.) Proceedings of the 6th International Workshop on Consuming Linked Data co-located with 14th International Semantic Web Conference (ISWC 2105), Bethlehem, Pennsylvania, US, October 12th, 2015, CEUR Workshop Proceedings, vol. 1426. CEUR-WS.org (2015)

18. Rudolph, S., Krötzsch, M., Hitzler, P.: All elephants are bigger than all mice. In: Baader, F., Lutz, C., Motik, B. (eds.) Proceedings of the 21st International Workshop on Description Logics (DL2008), Dresden, Germany, May 13–16, 2008, CEUR Workshop Proceedings, vol. 353. CEUR-WS.org (2008)

19. Steigmiller, A., Glimm, B., Liebig, T.: Reasoning with nominal schemas through absorption. J. Autom. Reasoning **53**(4), 351–405 (2014)

How to Keep a Reference Ontology Relevant to the Industry: A Case Study from the Smart Home

Laura Daniele[✉], Frank den Hartog, and Jasper Roes

TNO - Netherlands Organization for Applied Scientific Research, The Hague, The Netherlands
{Laura.Daniele,Frank.denHartog,Jasper.Roes}@tno.nl

Abstract. The Smart Appliance REFerence ontology (SAREF) is a shared model of consensus developed in close interaction with the industry to enable semantic interoperability for smart appliances. Smart appliances are intelligent and networked devices that accomplish some household functions, such as cleaning or cooking. This paper focuses on specific aspects of SAREF's development cycle, such as the design principles on which the ontology is based and the stakeholders' requirements from which certain modelling decisions originated. Moreover, we discuss the work to be done in the immediate future for SAREF to evolve concerning its maintenance, versioning, extension and governance. Open questions include how to guarantee the correct usage of SAREF, how to systematically manage the growth of extensions and specializations of SAREF in a consistent network of ontologies, and who should be responsible for these activities.

Keywords: Smart appliances · Ontologies · Semantic interoperability · Maintenance · Extension · Governance

1 Introduction

The Smart Appliance REFerence ontology (SAREF) is a shared model of consensus developed in close interaction with the industry to facilitate the matching of existing assets (i.e., standards, data models, protocols, specifications) in the smart appliances domain. Smart appliances are intelligent and networked devices that accomplish some household functions, such as cleaning or cooking. Smart appliances play an active role in the energy management of the buildings and are of strategic importance in achieving the goal of higher energy efficiency in the European economy [1]. SAREF focuses on the concept of "device", which is a tangible object designed to accomplish a particular task in households, common public buildings or offices. In order to accomplish this task, a device performs one or more "functions". For example, a washing machine is designed to wash (task) and to accomplish this task it performs, among others, a start and stop function. When connected to a network, a device offers a "service", which is a representation of a function to a network that makes the function discoverable, registerable and remotely controllable by other devices in the network. A device is also characterized by an "energy profile" and a "power profile" that can be used to optimize the energy efficiency in a home or office that are part of the building. SAREF is expressed in OWL-DL and contains 124 classes, 56 object properties and 28 datatype properties.

© Springer International Publishing Switzerland 2016
V. Tamma et al. (Eds.): OWLED 2015, LNCS 9557, pp 117–123, 2016.
DOI: 10.1007/978-3-319-33245-1_12

The documentation of SAREF is available online[1] where the ontology file can be downloaded in Turtle format[2]. A detailed description of the main classes and properties of SAREF can be found in our earlier work [3], in which we also described the approach taken during the SAREF development and we articulated the general principles learned from our experience in working with the community of industrial practitioners in the ontology creation process.

In this paper we focus on specific aspects of SAREF's development cycle, such as the stakeholders' requirements, the fundamental design principles used to create the ontology and the best practices followed to describe, document and publish SAREF in order to promote its usage in the smart appliances community. Moreover, we discuss the work to be done in the immediate future for SAREF to evolve concerning its maintenance, versioning, extension and governance. Open questions are how to guarantee the correct usage of SAREF, how to systematically organize future extensions and specializations in a consistent network of ontologies, and who should be responsible for these activities. The paper is structured as follows. The next section addresses various aspects that help to understand the modelling decisions underlying SAREF (i.e., requirements, principles and best practices), followed by the discussion on the maintenance, versioning, extension and governance of the ontology network sprouting from SAREF. The final section concludes the paper.

2   SAREF: Requirements, Principles and Best Practices

One of the most important stakeholders of SAREF is the European Commission (EC). The lack of a shared model of consensus in the smart appliances industry is a bottleneck for the EC to reach their energy saving ambitions [2, 4, 5]. Therefore the EC financed the development of the initial version of SAREF[3] [2], which project we subsequently executed[4]. During the development of SAREF, we had to take into account various requirements that the EC specifically mentioned in the tender document [2], and additional requirements set by our industrial collaborators. Often this meant that we had to balance apparently contradicting requirements. For example, the tender specification indicated OWL as the preferred ontology language for the project. However, the tender also stated that "the tenderer is free to suggest in his offer other tools or formalized languages, especially if they could facilitate collaborative aspects of ontology development and dynamic evolution of ontology networks in distributed environments". We decided to use OWL-DL as suggested, since it provided us with formal semantics and allowed a sufficient degree of semantic reasoning, being supported by a large number of software reasoning and consistency checking tools. Moreover, the tender required SAREF to be as complete as possible "to cover the needs of all appliances relevant for energy efficiency", and at the same time optimize the ontology "to be synthetic, compact

---

[1] http://ontology.tno.nl/saref.

[2] http://ontology.tno.nl/saref.ttl.

[3] http://ec.europa.eu/digital-agenda/en/news/invitation-tender-study-available-semantics-assets-interoperability-smart-appliances-mapping.

[4] https://sites.google.com/site/smartappliancesproject/home.

and with the minimum redundancy". The reference ontology resulting from our work was also meant to cover all semantic requirements as discovered in the project, but also "designed in a way that it can be expanded to cover future intelligence requirements". Finally, from a foundational point of view, we wanted to have a well axiomatized ontology to exploit the reasoning capabilities offered by OWL-DL, although the tender stated that the ontology under development was expected to be "a rather simple ontology as compared of state of the art ontology engineering level of complexity". In conclusion, it was required to produce an artefact with an exhaustive coverage in which all the different stakeholders in the domain could recognize their work, but which should be also rather simple, understandable and easy to use, in order to ease the adoption by the stakeholders. This was not a trivial task, and as described in [3] we had to compromise between the conceptual thinking underlying the world of formal ontologies and the more practical point of view of industrial stakeholders.

We created SAREF using fundamental principles of ontology engineering [6], such as reuse and alignment of concepts that are defined elsewhere. Since a large amount of work was already being done in the smart appliances domain, we have not invented anything new, but harmonized and aligned what was already there. The SAREF ontology was therefore built using core concepts that describe existing semantic assets (i.e., standards, data models, protocols, specifications) in the smart appliances community. A mapping that aligns concepts from existing semantic assets to core concepts in SAREF is available at the project website[5] as a complementary file to the project's final deliverable [7]. Moreover, SAREF reuses ontologies and vocabularies that have been developed in the Semantic Web community. In particular, SAREF directly imports the W3C WGS84 geo positioning vocabulary[6] and the W3C Time ontology[7]. SAREF also refers to the Ontology of units of Measure (OM)[8] to define the members of its "unit of measure" subclasses, such as the "power unit" class in the following example:

```
saref:PowerUnit
  rdf:type owl:Class ;
  rdfs:label "Power unit"^^xsd:string ;
  rdfs:subClassOf saref:UnitOfMeasure ;
  owl:oneOf (
     om:kilowatt
     om:watt
   ) ;
.
```

We have not directly imported OM since we only needed a reference to some basic units of measure and not the entire reasoning capability of this complex ontology, which in our opinion, if imported, would have confused the smart appliances industry - main

---

[5] D-S4 - SMART 2013-0077 - Smart Appliances - Mapping SAREF to short list assets: https://sites.google.com/site/smartappliancesproject/documents.
[6] http://www.w3.org/2003/01/geo/wgs84_pos.
[7] http://www.w3.org/TR/owl-time.
[8] http://www.wurvoc.org/vocabularies/om-1.8.

user of SAREF - who is rather pragmatic and not acquainted with (complex) ontologies. In contrast, we decided to include the W3C WGS84 geo positioning vocabulary and the W3C Time ontology as direct imports to fully exploit their reasoning capability and since they are reasonably small and usable also by non-expert users.

SAREF is furthermore based on the principle of modularity to allow separation and recombination of different parts of the ontology depending on specific needs. Towards this aim, SAREF provides building blocks that can be combined to accommodate different needs and points of view. The starting point in SAREF is the concept of "device". For example, a "switch" is a device. A device is always designed to accomplish one or more functions. Therefore, SAREF offers a lists of basic functions that can eventually be combined in order to have more complex functions in a single device. For example, the mentioned switch offers an actuating function of type "switching on/off". Each function has some associated commands, which can also be picked up as building blocks from a list. For example, the "switching on/off" function is associated with the commands "switch on", "switch off" and "toggle". Depending on the function(s) it accomplishes, a device can be found in a corresponding state. States are also listed as building blocks, making it easy and intuitive to combine devices, functions and states. The switch considered in our example can be found in one of the two states "on" or "off". SAREF also provides a list of properties that can be used to further specialize the functioning of a device. For example, a "light switch" specializes the more general "switch" for the purpose of controlling the "light" property.

According to best practices in the Semantic Web, SAREF includes basic metadata that allow others to correctly understand and properly reuse the ontology, such as creator, publisher, date of issue, title and description[9]. SAREF is self-descriptive since it contains labels, definitions and comments for its classes, and we also created a human-readable description that explains the main classes and properties. This description is available at the project website[10], together with the descriptions of the technology- and domain-specific ontologies that we used for the construction of SAREF[11]. We published the SAREF ontology at a stable URL (http://w3id.org/saref) in order to guarantee persistent access and facilitate its (re)usability in the smart appliances community.

3   Discussion

In this section we discuss the work to be done in the immediate future for SAREF to evolve and stay relevant to the EC and industry, also now that the initial project is completed. This work concerns the usage, extension, maintenance, versioning and governance of SAREF. The observations follow from our experience in the development of SAREF, but can be generalized for other (networks of) ontologies in different application domains. First, the adoption of SAREF and its correct usage need to be promoted. During the project, the EC and ETSI organized a number of workshops to allow us to

---

[9] From the Dublin Core Metadata Initiative (DCMI) Metadata Terms: http://purl.org/dc/terms.
[10] https://sites.google.com/site/smartappliancesproject/ontologies/reference-ontology.
[11] https://sites.google.com/site/smartappliancesproject/ontologies.

publicly present SAREF, answer the stakeholders' questions and collect their feedback in an interactive fashion. Version 1.0 of SAREF is available online since the end of the project and can be used by anybody that is interested in its capabilities. SAREF's online documentation is supported by the project's ad-interim reports[12], which are collected and harmonized in the final deliverable [7]. However, no explicit strategy has been defined on how to support users in using SAREF, especially if they only need a limited set of capabilities that suit their purpose, instead of the entire model. This issue becomes even more relevant when the interested users are not ontology experts, and therefore not acquainted to create new ontologies by importing or referencing to concepts defined elsewhere. We therefore conclude that there is a need for flexible and user-friendly solutions in terms of tools, methods, guidelines and best practices to (1) further promote the adoption of SAREF, and (2) allow third-party developers and users to utilize the ontology, either the whole model or a relevant subset of it.

Another important observation comes from the fact that SAREF is a conceptual artefact that does not cease to live because the project in which it was conceived is finished. SAREF has its own lifecycle and we expect it to continuously evolve in the future to cover new domains that are relevant in the home environment, such as e-health and entertainment. There will be new devices and consequently new functions, commands, services and so forth to be added as extensions to SAREF. Furthermore, it would be extremely beneficial to enhance the reasoning capabilities of SAREF by enabling its extension by means of rules. Moreover, the current concepts in SAREF are rather high-level, since SAREF contains core concepts that are recurring in the smart appliances domain. These concepts need to be specialized into a finer-grained level of detail to accommodate the requirements of specific use cases that come at hand. As a consequence, we envision that SAREF will grow into a network of ontologies that will require some governance. We should therefore investigate a modular and consistent way to enable this growth, while acknowledging that different stakeholders should be able to specialize the SAREF concepts according to their needs and points of view, add more specific relationships and axioms to refine the general (common) semantics expressed in the reference ontology, and create new concepts linked to existing concepts in SAREF. The minimum requirement is therefore that any extension/specialization must import or refer to SAREF.

An example on how to extend SAREF as described above is provided by an interoperability initiative taken by the EEBus[13] and Energy@home[14] associations. The initiative includes the development of an extension of SAREF to bridge the semantic gap between the EEBus and Energy@home data models. These two data models focus on the same concept, namely the concept of "power profile", but they use different terminologies. A power profile contains the information about the consumption and production of smart appliances in the household, for example a washing machine or a refrigerator. This information is exchanged between the appliances and the Consumer Energy Manager (CEM) for the purpose of energy efficiency optimization. SAREF already contains the concept of

---

[12] https://sites.google.com/site/smartappliancesproject/deliverables.

[13] http://www.eebus.org/en.

[14] http://www.energy-home.it.

power profile, but this concept needs to be specialized in more detail in order to accommodate the specific requirements of the use cases prescribed by EEBus and Energy@home. This initiative helps with creating the envisioned network of ontologies, learning by experience from these use cases, and laying out (initial) best practices that can be reused and improved by anybody interested in this task.

An additional and essential point of discussion concerns who should be responsible for the extension of SAREF, its maintenance and, more in general, the governance of the envisioned network of ontologies. Not only SAREF, but also the network of ontologies needs to be maintained in order to identify and correct defects, accommodate new requirements, and cope with changes over time. In principle, anybody can create a new ontology that makes use of SAREF and the creator of such ontology is responsible for its maintenance and versioning, independent from SAREF. In this way, the maintenance of the network will be distributed over the creators of the new ontologies sprouting from SAREF. To avoid inconsistency and confusion, we in contrast believe that the maintenance of SAREF should be delegated to a single party (e.g., an individual organization or a group of organizations) who should also take care of aligning SAREF with new ontologies in the network when necessary. Consider, for example, the case in which several extensions, like the one created for EEBus and Energy@home, are developed to accommodate different use cases, but present common recurring concepts or properties that could be "promoted" as core-upper concepts in SAREF. Which organization is going to implement the necessary updates and create the new version of SAREF? TNO could be a natural candidate as the creator of the first version of SAREF, but it is the EC who is the official owner of SAREF, ETSI is given the responsibility to adopt SAREF as a Technical Specification (TS) [8], and a large number of industrial stakeholders have closely collaborated as domain experts in SAREF's development. Also new parties such as W3C could play a role.

4   Conclusions

This paper follows our earlier work [3] in which we presented SAREF, the Smart Appliances REFerence ontology, pointing out the lesson learned from our collaboration with the industry during the ontology development process. In this paper we elaborated on several aspects of the development lifecycle of SAREF. We discussed various requirements that we had to take into account when creating SAREF, partly mentioned by the EC in the tender document, but also set by our industrial collaborators during the development process. Often this meant that we had to balance apparently contradicting requirements. We furthermore discussed how SAREF is based on fundamental principles of ontology engineering, such as reuse and modularity. We also pointed out best practices that we have used to describe, document and publish SAREF in order to guarantee persistent access and facilitate its (re)usability in the smart appliances community. We finally discussed the work to be done in the immediate future for SAREF to evolve concerning its maintenance, versioning, extension and governance. We concluded that there is a need for flexible and user-friendly solutions to further promote the adoption of SAREF and allow third-party developers and users to utilize the ontology, either the whole model or a relevant subset of it. Moreover, we identified the need for suitable mechanisms to define the SAREF extension and maintenance workflow, manage SAREF extensions and changes submitted by its users, possibly exploiting state-of-the-art versioning and consistency checking. The question of which organization(s) should be responsible for the maintenance, extension and governance of the SAREF network of ontologies remains open and needs special attention in the immediate future.

Acknowledgments. This work has been partly funded by the European Commission under contract number 30-CE-0610154/00-11.

References

1. Mertens, R., et al.: Manual for Statistics on Energy Consumption in Households. Publications Office of the European Union, Luxembourg (2013)
2. European commission: invitation to tender - study on the available semantics assets for the interoperability of smart appliances. Mapping into a Common Ontology as a M2 M Application Layer Semantics - SMART 2013/0077
3. Daniele, L., den Hartog, F., Roes, J.: Created in close interaction with the industry: the smart appliances reference (SAREF) ontology. In: Cuel, R., Young, R. (eds.) FOMI 2015. LNBIP, vol. 225, pp. 100–112. Springer, Heidelberg (2015)
4. ICT for a low carbon economy, eebuilding data models, energy efficiency vocabularies & ontologies. In: Segovia, R. (ed.) Proceedings of the 4th Workshop Organised by the EEB Data Models Community ICT for Sustainable Places, Nice, France, 9th–11th September 2013. European Commission, Brussels (2014)
5. den Hartog, F., Daniele, L., Roes, J.: Toward semantic interoperability of energy using and producing appliances in residential environments. In: 12th Annual IEEE Consumer Communications & Networking Conference (CCNC 2015), Las Vegas, USA, pp. 162–167. IEEE Press (2015)
6. Uschold, M., Gruninger, M.: Ontologies: principles, methods and applications. Knowl. Eng. Rev. **11**(2), 93–136 (1996). Cambridge University Press
7. Daniele, L., den Hartog, F., Roes, J.: Study on Semantic Assets for Smart Appliances Interoperability, D-S4 Final Report and D-S4 Mapping SAREF to Short List Assets. European Commission, Brussels (2015)
8. ETSI TS 103 264 V1.1.1, SmartM2M Smart Appliances Reference Ontology and oneM2M Mapping, ETSI (2015). http://www.etsi.org/deliver/etsi_ts/103200_103299/103264/01.01.01_60/ts_103264v010101p.pdf

An INSPIRE-Based Vocabulary
for the Publication of Agricultural Linked Data

Raúl Palma[1]([⊠]), Tomas Reznik[2], Miguel Esbrí[3], Karel Charvat[4],
and Cezary Mazurek[1]

[1] Poznan Supercomputing and Networking Center, Poznań, Poland
{rpalma,mazurek}@man.poznan.pl
[2] Masaryk University, Brno, Czech Republic
tomas.reznik@sci.muni.cz
[3] Atos, Madrid, Spain
miguel.esbri@atos.net
[4] Wirelessinfo, Litovel, Czech Republic
charvat@ccss.cz

Abstract. FOODIE project aims at building an open and interoperable agricultural specialized platform on the cloud for the management, discovery and large-scale integration of data relevant for farming production. In particular, the integration focuses on existing open datasets as well as their publication in Linked data format in order to maximize their reusability and enable the exploitation of the extra knowledge derived from the generated links. Based on such data, for instance, FOODIE platform aims at providing high-value applications and services supporting the planning and decision-making processes of different stakeholders related to the agricultural domain. The keystone for data integration is FOODIE data model, which has been defined by reusing and extending current standards and best practices, including data specifications from the INSPIRE directive which are in turn based on the ISO/OGC standards for geographical information. However, as these data specifications are available as XML documents, the first step to publish Linked Data required transforming or lifting FOODIE data model into semantic format. In this paper, we describe this process, which was conducted semi-automatically by reusing existing tools, and adhering to the mapping rules for transforming geographic information UML models to OWL ontologies defined by the ISO 19150-2 standard. We describe the challenges associated to this transformation, and finally, we describe the generated ontology, providing an INSPIRE-based vocabulary for the publication of Agricultural Linked Data.

1   Introduction

The agriculture sector has been of strategic importance for both European citizens (consumers) and European economy (regional and global) since the conception of the EU; it was one of the first sectors of the economy to receive the attention of EU policymakers [3]. And despite the fact that its contribution to

© Springer International Publishing Switzerland 2016
V. Tamma et al. (Eds.): OWLED 2015, LNCS 9557, pp. 124–133, 2016.
DOI: 10.1007/978-3-319-33245-1_13

the overall EU economy has slightly decreased during the previous years, agriculture (with forestry and fishing) represented about 1,7 % of the EU-28 Gross Added Value and accounted for 4.9 % of the total number of persons employed in 2013 [4]. As a result, the EU has developed policies and innovation programs that tackle the challenges associated to improve the efficiency of agricultural activities with a limited environmental footprint (see [2]).

Along these lines, we claim that in order to make economically and environmentally sound decisions, the different stakeholders groups involved in the agricultural activities need integrated access to multiple and heterogeneous sources of information collected by multiple applications and devices. In this context, FOODIE project[1] aims at building an open and interoperable cloud-based platform addressing among others the integration of data relevant to farming production, particularly from open datasets, as well as their publication in Linked data format.

In order to build such platform, we defined the modeling approach for the categories of information the platform will have to deal with, including their thematic, spatial and temporal characteristics as well as their meta-information. Such approach relies on reusing and extending standards and best practices to specify FOODIE data model. In particular we reused data specifications from the INSPIRE directive[2], which in turn are based on ISO/OGC standards for geospatial services and formats[3], thus applying the ISO/OGC-approach of modeling physical things, so-called "features". The specifications are defined as UML models and are available in different XML-based formats (e.g., GML, XMI) and as Enterprise Architect (EA)[4] projects. Accordingly, FOODIE data model was specified in UML by extending and specializing INSPIRE data model for Agricultural and Aquaculture Facilities (AF) [7].

However, according to the methodological guidelines for Linked Data publication [6], we need to specify the model for the representation of the data objects and their relationships. This usually involves the specification of a lightweight ontology (or vocabulary), reusing standard vocabularies wherever possible. In our case, this required transforming FOODIE UML data model into a lightweight ontology. But in addition to reusing standard vocabularies, our requirement was to comply with standard rules for mapping ISO UML models to OWL ontologies. In this paper we describe this transformation process, and the resulting ontology that can be used to represent agricultural-related data in a Linked Data representation or serialization (e.g., Turtle, RDF/XML).

2   Transformation

Based on the evaluation of different approaches in the literature for similar transformation process (see Sect. 4), we decided to follow a semi-automatic one using

---

[1] http://foodie-project.eu/.

[2] INSPIRE directive (http://inspire.ec.europa.eu/) aims at building a Pan-European spatial data infrastructure (SDI), requiring EU Member States to make available spatial data, from multiple thematic areas, according to established implementing rules using appropriate services [5].

[3] http://www.opengeospatial.org/standards/is.

[4] http://www.sparxsystems.com/.

ShapeChange tool. ShapeChange can process application schemas for geographic information from a UML model (e.g., XMI) and derive implementation representations, such as XML schemas, feature catalogs, and ontologies. In our case, we were interested in the OWL processor that is based on the ISO 19150-2 standard [8] defining rules for mapping ISO geographic information UML models to OWL ontologies.

2.1   Pre-processing Tasks

Source Model. FOODIE UML data model required some changes before processing it in ShapeChange. These changes led to the release of a new version (v4.3.2)[5] and include: (i) assignment of INSPIRE application schema stereotype to include the target namespace; (ii) fixing inconsistent range usage for attribute *code*; (iii) naming target sides of aggregations and associations for the generation of named object properties. The model was then published as XMI from EA tool, but we had to remove manually the ASCII code for Carriage Return encoded as an XML character reference in the file.

ShapeChange Configuration. The primary mechanism for providing arguments to ShapeChange is the configuration file. The two main components of this file are the encoding rules and the mappings. The first drives (broadly) the conversion from an application schema in UML to another data structure. The second supports customized mappings from UML classes to target OWL elements, by enabling the specification of generic rules. Additionally, the ShapeChange processor relies on different base ontologies for the generation of the RDF model, and thus the configuration file includes namespaces definitions for these ontologies. In particular, the processor uses geo-spatial ontologies, including those based on ISO 19100 series standards GeoSPARQL OGC standard and INSPIRE specifications; and in line with the Linked Data publication guidelines, it reuses several standard vocabularies like rdf, skos, dublin core and PROV.

We used as starting point the sample configuration settings in http://shapechange.net/targets/ontology/uml-rdfowl-19150-2/ and customized it according to our needs[6]. In particular, we applied the following rules: (i) ontologies are created only for the selected schema; (ii) constraints on properties and classes are specified; (iii) feature types get a subClassOf declaration to the GeoSPARQL FeatureType class; (iv) feature types get a subClassOf declaration to the ISO 19150-2 FeatureType and ISO 19109 AnyFeature classes; (v) data types get a subClassOf declaration to the ISO 19150-2 Datatype class and code lists get a subClassOf declaration to ISO 19150-2 Codelist class; (vi) cardinality restrictions are specified; (vii) allValuesFrom restrictions are not specified; (viii) minCardinality is set to 0 for voidable properties; (ix) *dc:source* is included only on the ontology subject;

---

[5] Available at https://git.man.poznan.pl/stash/projects/FOOD/repos/model/browse/.

[6] https://git.man.poznan.pl/stash/projects/FOOD/repos/model/browse/shapechange-conf.

(x) association names are not specified; (xi) the namespace abbreviation for the application schema is used for the ontology name and filename.

We included more than ten mapping entries in the configuration file for the classes and properties referenced in the model (see Sect. 3). We also fixed and added several namespaces in the configuration file, i.e., many of the namespaces for the geo-spatial ontologies were outdated or incorrect (e.g., INSPIRE, iso19150-2 and iso19109 ontologies), and we needed to include new namespaces for the mapping entries we created (e.g., iso19103, iso19108 and iso19115-citation ontologies).

Base Ontologies. The base INSPIRE ontology (the schema for basic types used by multiple themes)[7] was slightly modified to load it correctly, namely we: (i) added namespace of geosparql ontology (missing); (ii) fixed namespace of iso19150-2 based ontology, and removed the ontology import statement (because of few inconsistencies - see discussion below); (iii) fixed *VerticalPositionValue* datatype declaration; (iv) changed the ontology namespace to avoid multiple base prefixes; (v) replaced range of object properties for which no RDF representation was known from *owl:Class* to *owl:Thing*.

The original iso19103 ontology[8] treated a few datatype as classes. For instance *Number* is defined as an equivalent class to the union of primitive numerical datatypes (*xsd:decimal*, *xsd:double*, *xsd:float* and *xsd:integer*), and as a result it was declared both as class and datatype. We removed the class declarations, however they are still being treated as classes (as it was intended). This is possible in RDF, but in OWL terms this means that we have an OWL full ontology, as in all reasonable profiles (OWL 2 DL and below) datatypes and classes need to be disjoint.

Overall, we found some issues with the ontologies based on the ISO 19100 series standards. They are in provisional state, although they were created between 2012 and 2013, and in many cases the versions changed drastically. For instance, ISO 19100 series standards define UML profiles that include a list of stereotypes and basic types to be used in application schemas. Accordingly, the ISO 19150-2 based ontology defined classes for these stereotypes, including ⟨⟨datatype⟩⟩, ⟨⟨codelist⟩⟩, ⟨⟨featureType⟩⟩, and the base class ⟨⟨anyFeature⟩⟩. However, the latest version of this ontology does not declare all these classes, as it did in the previous version (used in ShapeChange). Additionally, the ontologies miss several elements from the standard and in most cases the ontologies are only available as OWL full ontologies (e.g., treating datatypes as classes). We tried unsuccessfully to reach the developers to discuss these issues.

2.2   Post-processing Tasks

We had to make some manual fixes in the ontology after executing the transformation, including updating incorrect namespaces added automatically by the processor rules (hard-coded), adding missing prefixes and removing unnecessary

---

[7] http://portele.de/ont/inspire/base.ttl.

[8] http://def.seegrid.csiro.au/isotc211/iso19103/2005/basic.

imports of ontologies to avoid ending up with a heavy ontology. Additionally, as ShapeChange only processes the selected schema (i.e., FOODIE data model), we had to add manually the ontology elements (corresponding to the UML elements) of the base INSPIRE schemas, particularly those from the Agriculture and Aquaculture Facilities theme. Finally, we removed an axiom generated to constraint the cardinality of the property *rdfs:label* in a class expression (*rdfs:label* was the mapped property for the UML element "name") because *rdfs:label* is a predefined annotation property so it can only be used in annotations.

3   Ontology

FOODIE ontology is available at http://foodie-cloud.github.io/model/FOODIE. html, and a partial view of its taxonomy is depicted in Fig. 1. In the reminder we describe the main ontology elements (classes are in italics).

For the purposes of FOODIE, we found the lack of a feature on a more detailed level than *Site* that is already part of the INSPIRE AF data model. The main motivation was to represent a continuous area of agricultural land with one type of crop species, cultivated by one user in one farming mode (conventional vs. transitional vs. organic farming). Such concept is called *Plot* and represents the main element in the model, specially because it is the level to which the majority of agro data is related. One lower level than *Plot* is the *ManagementZone*, which enables a more precise description of the land characteristics in fine-grained areas. The *Plot* has associated two kinds of data: (i) metadata information, including properties: code (id), validity (when the plot started and ceased to exist), geometry (spatial extent), description and originType (manual, system); (ii) agro-related information, including:

- *ProductionType*, representing production-related data, comprising properties: productionDate (when the information was added/changed in the knowledge base (KB)); variety (assemblage of cultivated individuals that are distinguished by characteristics significant for agriculture, e.g., morphological, physiological, cytological, chemical); productionAmount (physical quantity of produced variety).
- *CropSpecies*, representing the planted crop species, comprising properties: date (when it started/ended to be planted on the Plot); cropArea (spatial extent on the Plot); cropSpecies (designation under which it is commonly known).
- *Alert*, representing alerts generated by the models integrated in the platform, comprising properties: code; type (according to user-defined classification, e.g. phytosanitary); description; checkedByUser (indication of user awareness); alertDate (creation); alertGeometry (spatial extent for which it is applicable).
- *Intervention*, representing the basic feature type for any application with explicitly defined geometry, comprising properties: type (free text (e.g., tillage, pruning)[9]); description; notes (user-defined); status (free text); creationDate

---

[9] In the cases of free text properties, it was not feasible to provide common code lists (e.g., values vary from country to country or from farm to farm).

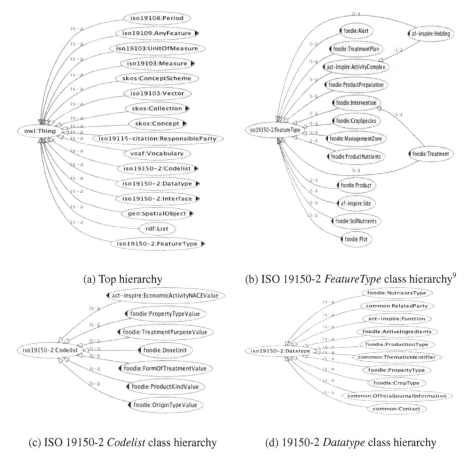

(a) Top hierarchy

(b) ISO 19150-2 *FeatureType* class hierarchy[9]

(c) ISO 19150-2 *Codelist* class hierarchy

(d) 19150-2 *Datatype* class hierarchy

Fig. 1. Partial view of FOODIE ontology taxonomy (same for ISO 19109 *AnyFeature* and geosparql *Feature* classes.)

(in the KB); interventionStart/End (when started/ended in the real word); interventionGeometry (spatial extent); supervisor (entity with authority to guarantee its execution); operator (person who executed it); evidenceParty (entity who added it in the KB); price.

The intervention has direct and indirect associations to the following entities (as depicted in Fig. 2[10]):

– *Treatment* comprising properties: quantity (applied physical quantity); tractorId (vehicle for machine applying it); machineId (machine applying it);

---

[10] In Fig. 2, boxes represent classes, arrows represent relations (black for subsumption, blue for direct relations), properties are listed inside the boxes (with a blue square for ObjectProperty and green square for DatatypeProperty).

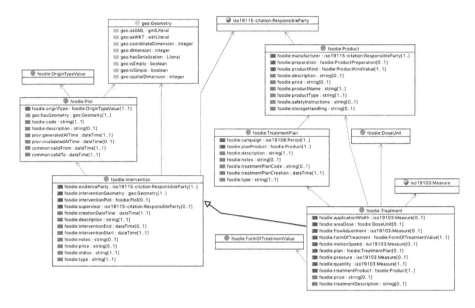

Fig. 2. Partial view of *Intervention* class

motionSpeed (recommended speed for its application); pressure (recommended pressure for its application); flowAdjustment (indication if flow adjustment was needed for its application); applicationWidth (width in which a machine is capable to apply it); areaDose (maximum application rate); formOfTreatment (id of its application, e.g., manual, aerial, from a code list); treatmentPurpose (rationale why it was used, e.g., weed, pest, from a code list); treatmentDescription.

- *TreatmentPlan* comprising properties: treatmentPlanCode; description; type; campaign (period to which it was designed); treatmentPlanCreation (in the KB); notes.
- *ProductPreparation* comprising properties: productQuantity (physical quantity of the applied product); solventQuantity (physical quantity of solvent applied); safetyPeriod (when a dissolved product may be used).
- *Product*, comprising properties: productCode; productName; productType (free text); productSubType (detailed classification - as free text); productKind (origin, e.g., organic, mineral - from a code list); description; manufacturer; safetyInstructions; storageHandling (for safe storage); registrationCode (id according to the national or other relevant registration scheme); registerUrl (link to the national (or other) registry); nutrients (id of nutrients, i.e., chemical elements and compounds necessary for plant growth, represented by *NutrientsType* class comprising properties for the amount of nitrogen, phosphorus pentoxide, potassium oxide and other chemical elements.
- *ActiveIngredients* with properties: code, ingredientName, and ingredient Amount.

4   Related Work

We identified and analyzed some relevant works that tackle the transformation of UML models into an OWL ontology, particularly those related to the geo-spatial domain. In [9], the authors propose a general approach for translating INSPIRE-compliant GML data models into OWL ontologies They consider the common characteristics that INSPIRE UML (and derived GML) models have with OWL ontologies to derive a set of general conversion rules and some ontological refinements for frequently used element types. The execution of the UML-OWL conversion, however, is not described and there is already a standard defining mapping rules for transforming geographic information UML models to OWL ontologies. In [10], the authors propose to generate an OWL ontology from their UML model with ShapeChange tool. They then propose the usage of annotations in the UML attributes with special meaning in RDF to use existing RDF vocabularies in the generated ontology. The latest version of ShapeChange, however, enables the specification of such mappings in their configuration files. Another relevant work is presented in [1] where the authors present a methodology for exposing INSPIRE data and metadata on the Semantic Web through GeoSPARQL endpoints. They follow a data-centric approach based on the usage of XSL Transformations to map INSPIRE-compliant metadata records and data elements from INSPIRE-compliant features into suitable RDF statements. In our case, we are more interested in a model-based approach where the resulting ontology can later be reused and extended, and where different source datasets can either stay in the original (geospatial) database (using virtual SPARQL endpoint) or lifted into a semantic store.

It is also worth mentioning that there exist some relevant ontologies and vocabularies in the agricultural domain, being AGROVOC the most notable. AGROVOC[11] is a multilingual controlled vocabulary from FAO covering areas like food, nutrition, agriculture, fisheries, forestry and the environment. Complementary FAO also developed agrontology[12], an OWL vocabulary providing a set of domain properties to AGROVOC. Another example is the Organic Agriculture ontology developed within the Organic.Lingua project[13] to enhance an educational Web portal[14] with content on Organic Agriculture and Agroecology. These resources are good examples of (multi-)domain ontologies which are applied particularly for the tasks of indexing, annotation, and retrieval of resources. In our case, however, we are interested in an application oriented ontology that can deal with all the categories of information the FOODIE platform will have to deal with, including farming tasks and activities, and their publication as Linked Data. Nevertheless, we plan to align our ontology with concepts from some of these resources, particularly from AGROVOC.

---

[11] http://aims.fao.org/vest-registry/vocabularies/agrovoc-multilingual-agricultural-thesaurus.
[12] https://aims-fao.atlassian.net/wiki/display/AGV/Agrontology.
[13] http://www.organic-lingua.eu/.
[14] http://www.organic-edunet.eu/en.

5   Conclusion

The publication of Agricultural Linked Data is unfortunately still not a common practice. One of the key issues for this is that although there are some vocabularies/ontologies for modeling the domain concepts and relations, there is still a lack of vocabularies for modeling all the relevant information within farm-oriented applications, including their tasks and activities, as Linked Data. In FOODIE project, we have developed an ontology for this task which complies and adheres to existing standards for the representation of geo-spatial data relevant for agriculture. In particular, we extended and specialized INSPIRE UML data model for Agricultural and Aquaculture Facilities and transformed this model into a lightweight ontology. We conducted this process semi-automatically reusing ShapeChange tool, which enables the transformation of UML models in XMI into OWL ontologies. The transformation required several pre and post processing tasks, in order to build the final ontology. We described in detail this process, the challenges associated, and finally, we presented the resulting ontology. At the moment we are working on the application of this ontology in FOODIE platform.

Acknowledgments. The research reported in this paper has been supported by the EU FOODIE project (http://foodie-project.eu/, CIP-ICT-PSP-2013-7, Pilot B no. 621074).

References

1. Alexakis, M., Athanasiou, S., Georgomanolis, N., Patroumpas, K., Stratiotis, T.: D2.7.1 geodata.gov.gr Geospatial Data as Linked Data. Technical report D2.7.1, ATHENA, June 2014. http://svn.aksw.org/projects/GeoKnow/Public/D2.7.1_Geodata.gov.gr_Geospatial_Data_as_Linked_Data.pdf
2. European Comission: Agricultural and rural development, research and innovation, challenges (2015). http://ec.europa.eu/agriculture/research-innovation/challenges/index_en.htm
3. Eurostat: Enlargement countries agriculture, forestry and fishing statistics (2014). http://goo.gl/Sm121D
4. Eurostat: Enlargement countries agriculture, forestry and fishing: tables and figures (2014). http://goo.gl/m3mkFA
5. Fichtinger, A.: INSPIRE Roadmap and Implementation. In: INSPIRE-GMES Information Brochure, 7th edn., pp. 2–10. Technische Universität München, October 2011
6. Hyland, B., Atemezing, G., Villazon-Terrazas, B.: Best practices for publishing linked data (2014). http://www.w3.org/TR/ld-bp/
7. INSPIRE Thematic WG Agricultural and Aquaculture Facilities. D2.8.III.9 data specification on agricultural and aquaculture facilities, December 2013. http://goo.gl/eWi6rq
8. ISO: Standard 19150–2:2015: Geographic information - Ontology - Part 2, July 2015. http://www.iso.org/iso/home/store/catalogue_tc/catalogue_detail.htm?csnumber=57466

9. Tschirner, S., Scherp, A., Staab, S.: Semantic access to inspire. In: Terra Cognita 2011 Workshop Foundations, Technologies and Applications of the Geospatial Web. Citeseer (2011)

10. Van den Brink, L., Janssen, P., Quak, W.: From geo-data to linked data: automated transformation from GML to RDF. Linked Open Data-Pilot Linked Open Data Nederland (2013)

Towards a Core Ontology of Occupational Safety and Health

Agnieszka Ławrynowicz$^{(\boxtimes)}$ and Ilona Ławniczak

Institute of Computing Science, Poznan University of Technology, Poznan, Poland
`alawrynowicz@cs.put.poznan.pl`

Abstract. We describe the core module of the Occupational Safety and Health Domain Ontology (OSHDO-Core) v1.0, we have developed. We also discuss the requirements specification and modeling and ontology engineering issues encountered during the design process. The resulting OSHDO-Core contains the core vocabulary of the domain with basic ontological distinctions. The ultimate goal is to establish a common vocabulary and a core formal model of the domain.

Keywords: Occupational Safety and Health · Ontology

1 Introduction

Occupational Safety and Health (OSH) is defined as the scientific domain dealing with the anticipation, identification, evaluation and control of *hazards* that emerge in or from the *workplace* and that may negatively impact the *health* and well-being of *workers* [1]. It is a broad, multidisciplinary domain since a variety of workplaces and hazards exist. *World Health Organization (WHO)* considers rather the discipline of *occupational health* defined as dealing with "all aspects of health and safety in the workplace and has a strong focus on primary prevention of hazards."[1] Health is considered as a state of complete well-being and not just the absence of a disease. Whereas the term *safety engineering* refers to a related engineering discipline whose goal is to ensure acceptable levels of safety provided by engineered systems. Studying the possible impact on the neighbouring communities and environment is also being considered as part of OSH.

Our contribution in this work is the specification of the core vocabulary of the domain with basic ontological distinctions. We have developed the core module of the *Occupational Safety and Health Domain Ontology (OSHDO-Core)* v1.0 that we discuss in this paper.

The rest of the paper is structured as follows. In Sect. 2 we discuss the requirements. In Sect. 3 we review related ontologies and non-ontological resources. In Sect. 4 we describe the content of OSHDO-Core. In Sect. 5 we provide a discussion and lessons learnt with respect to modeling issues and ontology engineering issues. In Sect. 6 we provide a summary and the future work agenda.

---

[1] http://www.wpro.who.int/topics/occupational_health/en/.

© Springer International Publishing Switzerland 2016
V. Tamma et al. (Eds.): OWLED 2015, LNCS 9557, pp. 134–142, 2016.
DOI: 10.1007/978-3-319-33245-1_14

2   Requirements Specification

We have identified several use cases for OSHDO[2], such as: analyzing occupational risks, analysing and making predictions concerning the causes and consequences of accidents or potentially accidental events, developing training materials, linking workers with events that occurred in the company, creating annual statistics. We have also identified the possible users of the ontology: the worker, the employer, the occupational safety and health specialist, general inspectors (e.g., national labour inspectorate), the (central) statistical office.

Below we present sample competency questions. The questions are organized into thematic groups according to Ishikawa diagrams [2] (also called fishbone diagrams or cause–and–effect diagrams). This technique is commonly used to locate causes of a mistake, accident, or potentially accidental events.

CQ1 Man:

- What is an unsafe act?
- What regulates training of workers?
- Can mental and physical condition cause accidents at work?

CQ2 Method:

- To which group of causes belongs inadequate adaptation of technology to the work process?
- What describes the procedures to be followed at work?

CQ3 Management:

- To what group of accident causes belongs inappropriate distribution and storage of objects of labor (raw materials, semi-finished products, etc.)?
- What causes the division of work or task planning to be incorrect?

CQ 4 Machine:

- Which group of causes includes excessive exploitation of material?
- Which group of causes includes improper repairs and renovations?

CQ 5 Material:

- What are hidden causes of material defects?
- Does incorrectly protected material can cause near miss event?

CQ 6 Environment:

- What are the unsafety conditions?
- What group of hazards affects the environment?
- What factors belong to a group of chemical/physical hazards?

---

[2] The complete list may be found at http://semantic.cs.put.poznan.pl/ontologies/oshdo/.

CQ 7 Risk assessment:

- What is the most common cause of accidents at work?
- What is WEEL (Workplace Environmental Exposure Level)?
- What methods can be used to minimize occupational risk?
- What threat is most common in occupational risk assessment?
- What are the most commonly used means of corrective or preventive actions?
- Which means are part of the PPE (personal protective equipment)?
- Which groups of threats/hazards can be distinguished ?
- What kind of health effects are reversible?
- What kind of a situation/an event is that in which there nearly happened an accident at work?
- What is the severity of the consequences defined as resulting from injuries and diseases (e.g. third-degree burns, occupational hearing loss)?

The complete set of the questions may be found at http://semantic.cs.put. poznan.pl/ontologies/oshdo/.

3   Related Work

The occupational safety and health overlaps with a number of disciplines including safety engineering, (occupational) medicine, psychology, epidemiology, human factors and ergonomics, physiotherapy and rehabilitation, law and others. The ontologies for many of those fields exist, but focused on the particular domain. Below, we briefly review the ontologies from the overlapping domains.

The major terms in OSH are hazard (threat) and risk. Threat is considered in Internet and/or Cyber Security, and in [3] an ontology is described for the latter domain, but it is modeled with a different focus and scope. The Risk base ontology from the EU project RISCOSS [4] re-uses concepts from another risk ontology, defined in the EU project Musing[3], definitions from standards such as ISO 31000, and terms defined in risk management methodologies. But it does not include a concept of hazard. A hackathon at Ontology Summit 2014[4] was devoted to modeling hazard and risk related concepts but in the travel domain.

There are many ontologies that might be considered relevant for modeling hazards, such as from the life sciences domain [5] which include knowledge on potential hazards - biological and chemical agents - and diseases. They may be useful in further steps for modeling OSHDO branches beyond the core module.

Legal ontologies such as LKIF-Core Ontology [6] describe the issues related to law, legislation, norms, but none of them models vocabulary typical to OSH (specific to safety management, as how to implement the duties related to safety, what are the requirements for carrying out activities aimed at risk reduction etc.). Similarly, ontologies related to management or business processes [7] do not contain key concepts from the OSH domain, e.g. to describe the consequences

---

[3] http://cordis.europa.eu/ist/kct/musingsynopsis.htm.
[4] http://ontolog.cim3.net/OntologySummit/2014/.

of impacts in the labor process, prevention methods, procedures and security measures which must be used to deal with hazardous substances and situations.

Identifying and analyzing the available ontologies revealed lack of one that fully covers the scope and vocabulary for the subject of OSH. Thus we aim at creating a new ontology on the basis of available resources.

4   Overview of the Content of OHSDO-Core

4.1   Main Concepts of OHSDO-Core

To define a set of core concepts we have examined non-ontological resources including: the taxonomy of categories of OSHwiki[5], definitions from standards such as OHSAS 18001 as well as glossaries of terms[6]. We have also studied handbooks on the domain [1,8] and the definitions therein. Below, we describe basic terms of OHSDO-Core v1.0. The ontology may be downloaded from http://semantic.cs.put.poznan.pl/ontologies/oshdo/OSHDO-Core.owl.

OccupationalSafety and Health are two important terms in the domain that are often misinterpreted. For the layperson, 'safety' is to not getting injured. But professionals operate by referring to the likelihood or risk of an occurence of such an event. Safety is understood as "operating within an acceptable or low probability of risk" [8] that concerns potential harm to people and non-human resources of the enterprise (equipment and facilities). Similarly, Health is commonly interpreted as the absence of a disease. But in the OSH domain, a broader definition, as the one of WHO (see: Sect. 1), is adapted.

OccupationalHazard is the major term in the domain. It is defined as a *potential source* of Harm (Injury or other health deterioration) on someone or something. It is a *threat* that may be caused by a source, situation, or act with a potential for harm. It may be responsible for Incidents at Workplace and/or occupational diseases (OccupationalDisease) under occurrence of a certain HazardousEvent or condition. HazardousEvent is as an event where at least one participating Worker is exposed to an OccupationalHazard. HazardousEvent may casually follow some Cause and may cause one or more Consequence. OccupationalExposure is a measure of the extent (a dose) to which a Worker (or Equipment) is likely to be exposed to - or may be influenced by - the OccupationalHazard. HazardousSituation is a situation that participates in one or more HazardousEvents (Hazardous Situation ontology design pattern (ODP) is depicted in Sect. 4.2). It represents a 'snapshot', a setting that is associated with a HazardousEvent. There is a variety of hazards e.g., physical hazards, electrical hazards, radiation hazards, noise, and vibration. HazardIdentification belongs to main areas of OSH.

Risk, in the simplest case, is the likelihood that Harm will actually occur ($Risk = Hazard \times Exposure$). A broader meaning of Risk also considers the likelihood of the amount of Harm that the exposure to a hazard will cause

---

[5] http://oshwiki.eu.

[6] http://www.iapa.ca/pdf/iapa_glossary.pdf.

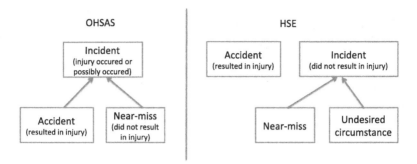

Fig. 1. Accident as an event of type of Incident (according to OHSAS versus Accident and Incident as different types of events (according to HSE (http://www.hse.gov.uk)).

(Severity). For instance, in [1] there is provided the following formula to define Risk: $Risk = Probability \times Severity$. The OHSAS 18001:2007 standard defines risk as "combination of the likelihood of an occurrence of a hazardous event or exposure(s) and the severity of injury or ill health that can be caused by the event, accident(s) or exposure(s)".

RiskAssessment concerns analyzing the risks and determining the level of acceptability of risk, based on the reviews of the risk including risk analysis and risk evaluation. It shall consider acceptable risks, for which changes are not necessary and those for which actions are needed to be taken to reduce Risk. Action is a short-term event. PreventiveAction and CorrectiveAction may be distinguished (for preventing incidents and eliminating similar incidents to those that has just happened).

Harm occurs in practice when there are both: the OccupationalHazard and OccupationalExposure. It may be a result of Incident or Accident at Workplace causing negative consequences for the life and health of a Worker or Equipment. OccupationalDisease is primarily a result of OccupationalExposure to risk factors in the Workplace arising from specificity of the tasks that Worker performs.

Various definitions of Incident and Accident, regarded as unintended events at work, exist in the literature (see: Fig. 1). We adopt the definition of Incident from OHSAS 18001 standard, where Incident is regarded as "a work-related event(s) in which an injury or ill health (regardless of severity) or fatality occurred, or could have occurred"[7]. Those events are nearly always preceded by unsafe acts of workers, hazardous conditions in the workplace, or both.

Worker is a physical person employed by an employer to carry out the tasks laid down in the contract of employment. Employer is a physical or legal person that operates under law with regard to an employment relationship with the Worker. From a safety standpoint Employer is responsible for the condition of the company as well as for activities that must be performed in order to identify and minimize risks. Workplace is the physical location and its surroundings where the Worker takes actions related to performing any work that is

---

[7] http://www.ohsas-18001-occupational-health-and-safety.com.

organized by the management and subject to ongoing monitoring. Workplace also includes machines (Machine), equipments Equipment and tools needed to do the job. Hazardous environment at work is associated with unsafe conditions, i.e. such conditions, not directly caused by the action or inaction of one or more workers (e.g., faulty design) that may lead to an Incident or Injury if uncorrected.

OSH-Policy is defined by OHSAS 18001:2007 as "overall intentions and direction of an organization related to its OH&S performance as formally expressed by top management". It is a method for guiding carrying out actions. It states the principles and rules to guide actions. PreventionStrategy and ControlStrategy are both plans of actions such as, for instance, related to personal protective equipment, or workplace health promotion. OSH-Management is a geared effort of reducing the risk into acceptable threshold level, then keep it at the same or lower level. The safety management system OSH-ManagementSystem is regarded as a set of functions that are decisive in defining and implementing security policies in the workplace. OSH-Performance is a measurable, definite in time score indicating the level of the performance of the company in terms of OSH after the implementation of work safety management system. PreventiveMeasure and ProtectiveMeasure are used to assess, prevent and reduce occupational Risk. Efforts are made where risk assessment indicates an inadmissible level of risk to reduce it and improve the Workplace.

4.2   The Hazardous Situation Ontology Design Pattern

We briefly present an ontology design pattern we have developed to represent hazardous situations and events. It is intended to provide a building block for modeling hazardous situations, i.e. the situations where one or more objects is exposed to one or more hazards to some extent (exposure value). The pattern is depicted in Fig. 2. A longer description may be found in [9].

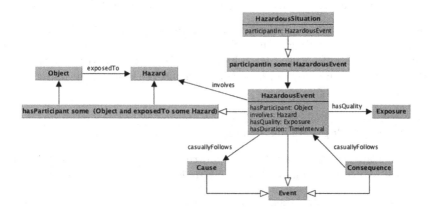

Fig. 2. The hazardous situation ontology design pattern

5   Discussion

5.1   Modeling Issues

The main difficulties we have encountered during modeling are related to the presence of various alternative definitions and interpretations of the core terms of OSH that may be found in the literature and within the proposed standards.

Risk. Risk is the term often being confused with OccupationalHazard. Some dictionaries give imprecise definitions or even combine those two terms (e.g. "a danger or risk"), and many people apply those terms interchangeably. But hazard is only one of the components of Risk. Several definitions and formulas have been proposed in the literature and it remains to be further researched what is the most proper formal definition in the context of OSH.

Hazard. During making modeling choices, we investigated the modeling pattern from the RISCOSS ontology [4]. RISCOSS models Risk, with exposure as a derived value, and RiskEvent with likelihood as a derived value. RiskEvent is an Event, and as such it corresponds most closely to Incident or HazardousEvent in OSHDO. The report from the hackathon at Ontology Summit 2014 uses the term hazard also as of an event. RiskEvent is a superclass of accidents in this view. Thus the term Factor from the report (such as EnvionmentalFactor, e.g. Limited_Sight) is closest to our interpretation of Hazard. Hazard in our case, is not an event. It is much closer in the meaning to Threat from [3].

Incident vs Accident. The definition of these terms is important in the context of prevention of these events. Although the term Incident is regarded increasingly as a broad term encompassing all events causing injury or material damages, also near-miss events, this is not always the case. Incident is often also referred to an event that have had the potential to cause harm, but didn't. Incident is then regarded as a synonym for a near-miss event. These differences in terminology and definitions have to be taken into account when reading OSH literature.

Hazardous Situation vs Hazardous Event. Before arriving in the Hazardous Situation ODP we had struggled in differentiating situations from events. In the literature, they are often used interchangeably. But we wanted to capture that by the situation we understand a 'setting', a 'context', more in the spirit as it is defined in the DOLCE ontology.

5.2   Ontology Engineering Issues

During this and previous ontology engineering efforts [10], we have identified needs with regard to ontology engineering tools that we summarize below:

- support for Ontology Design Patterns; this includes support for handling groups of axioms, rather than individual axioms, analogously to handling composite objects in vector graphics editors after 'grouping' operation to manipulate multiple axioms collectively;

- support for axiom generalisations, i.e. for templates of recurring modeling in axioms, where some entities in an axiom are replaced by variables (see for instance [11]). This should enable an ontology engineer to work at the level of the pattern and not that of the set of OWL axioms. Such templates could be used as macros to facilitate ontology population (see: Populous tool [12]);
- support for provenance-aware versioning, i.e. the use of a provenance and an ontology change vocabulary [13] to track entity provenance (who created what etc.), facilitate querying provenance, and 'rollback' operations.

6 Summary and Future Work

In this paper, we have presented the version v1.0 of the core module of the Occupational Safety and Health Domain Ontology (OSHDO-Core). We have described the main terms of the ontology and the Hazardous Situation ODP. We have discussed the modeling choices made and ontology engineering issues.

The future work agenda includes investigating, selecting and modeling further ontology design patterns that may be included into the ontology. We plan to further enhance the scope and the axiomatisation of the ontology.

Acknowledgements. This work was partially supported from the PARENT-BRIDGE program of Foundation for Polish Science, cofinanced from European Union, Regional Development Fund (Grant No POMOST/2013-7/8).

References

1. Alli, B.: Fundamental Principles of Occupational Health and Safety. International Labour Office (2008)
2. Ishikawa, K.: Guide to quality control. Industrial engineering and technology. Asian Productivity Organization (1976)
3. Oltramari, A., Cranor, L.F., Walls, R.J., McDaniel, P.: Building an ontology of cyber security. In: Proceedings of STIDS 2014 (2014)
4. Siena, A., Morandini, M., Susi, A.: Modelling risks in open source software component selection. In: Yu, E., Dobbie, G., Jarke, M., Purao, S. (eds.) ER 2014. LNCS, vol. 8824, pp. 335–348. Springer, Heidelberg (2014)
5. Whetzel, P.L., Noy, N.F., Shah, N.H., Alexander, P.R., Nyulas, C., Tudorache, T., Musen, M.A.: Bioportal: enhanced functionality via new web services from the national center for biomedical ontology to access and use ontologies in software applications. Nucleic Acids Res. **39**(suppl), W541–W545 (2011)
6. Hoekstra, R., Breuker, J., Di Bello, M., Boer, A.: LKIF core: principled ontology development for the legal domain. In: Proceedings of the 2009 Conference on Law, Ontologies and the Semantic Web: Channelling the Legal Information Flood, pp. 21–52. IOS Press, Amsterdam, The Netherlands (2009)
7. Gasevic, D., Guizzardi, G., Taveter, K., Wagner, G.: Vocabularies, ontologies, and rules for enterprise and business process modeling and management. Inf. Syst. **35**(4), 375–378 (2010)
8. Friend, M., Kohn, J.: Fundamentals of Occupational Safety and Health. Bernan Press, London (2014)

9. Lawrynowicz, A., Lawniczak, I.: The hazardous situation ontology design pattern. In: Proceedings of the 6th Workshop on Ontology and Semantic Web Patterns (WOP 2015) (2015)

10. Keet, C.M., Lawrynowicz, A., d'Amato, C., Kalousis, A., Nguyen, P., Palma, R., Stevens, R., Hilario, M.: The data mining optimization ontology. Web Semant. Sci. Serv. Agents World Wide Web **32**, 43–53 (2015)

11. Mikroyannidi, E., Iannone, L., Stevens, R., Rector, A.: Inspecting regularities in ontology design using clustering. In: Aroyo, L., Welty, C., Alani, H., Taylor, J., Bernstein, A., Kagal, L., Noy, N., Blomqvist, E. (eds.) ISWC 2011, Part I. LNCS, vol. 7031, pp. 438–453. Springer, Heidelberg (2011)

12. Jupp, S., Horridge, M., Iannone, L., Klein, J., Owen, S., Schanstra, J., Wolstencroft, K., Stevens, R.: Populous: a tool for building OWL ontologies from templates. BMC Bioinf. **13**(S–1), 1 (2012)

13. Palma, R., Haase, P., Corcho, Ó., Gómez-Pérez, A.: Change representation for OWL 2 ontologies. In: Proceedings of OWLED 2009 (2009)

Towards a Visual Notation for OWL: A Brief Summary of VOWL

Steffen Lohmann[1(✉)], Florian Haag[1], and Stefan Negru[2]

[1] Institute for Visualization and Interactive Systems (VIS), University of Stuttgart,
Universitätsstr. 38, 70569 Stuttgart, Germany
{steffen.lohmann,florian.haag}@vis.uni-stuttgart.de
[2] Faculty of Computer Science, Alexandru Ioan Cuza University,
Strada General Henri Mathias Berthelot 16, 700483 Iasi, Romania
stefan.negru@info.uaic.ro

Abstract. The Web Ontology Language (OWL) has no standardized visual notation in contrast to related modeling languages. However, the visual representation of individual and combined OWL elements as well as complete OWL ontologies can be very useful in many cases. We have developed the Visual Notation for OWL (VOWL) that defines graphical representations for most of the OWL language constructs. In contrast to related work, VOWL aims at a complete and well-specified notation that is easy to understand and implement. This paper reports on the current state of development and briefly describes the main design principles and considerations. At OWLED 2015, we conducted a special session to gather feedback on how to further improve the visual notation and to collect requirements for its future development.

1 Introduction

The Web Ontology Language (OWL) has become the 'lingua franca' for ontologies. Nearly all modern ontologies are modeled in OWL, and more and more OWL ontologies are developed every week. However, OWL is not a visual language. In contrast to related modeling languages that define visual notations, such as UML or ER diagrams, the OWL specifications do not include any recommendations on how to graphically represent the different OWL language constructs.

Yet, a visualization of individual and combined OWL elements as well as complete ontologies can be very useful in many situations. It can help in the development, exploration, verification, and sensemaking of ontologies. It can also be useful for teaching OWL, in order to illustrate the language constructs and possible combinations. Finally, visualizations of OWL are known to ease the communication between domain experts and ontology experts.

Although many visualizations for OWL ontologies have been developed in the past, only few define an explicit notation. Most visualizations use basic node-link diagrams or other visualization techniques to depict the concepts and relationships modeled in OWL [27,36], but neither provide any further description of the notation nor an explicit visual mapping for the individual OWL elements.

© Springer International Publishing Switzerland 2016
V. Tamma et al. (Eds.): OWLED 2015, LNCS 9557, pp. 143–153, 2016.
DOI: 10.1007/978-3-319-33245-1_15

We have developed the Visual Notation for OWL (VOWL) in an effort to close this gap and complement OWL with a well-specified visual notation that defines graphical representations for most language constructs. While we focused the development on an easy-to-understand notation for casual ontology users, VOWL can also be of use to ontology experts. A precise description of the visual notation can be found in the VOWL specification [40], while details on its implementation and evaluation are given in related publications [34–36]. Recently, we started work on version 3 of VOWL with the goals to incorporate the complete set of OWL language constructs and to further increase the scalability of the visual notation.

In this paper, we summarize the main design decisions, principles, and considerations related to the development of VOWL, and give an outlook on future directions. At OWLED 2015, we discussed the state of development in a special session, and collected feedback and requirements on how to further improve the visual notation.

2   Related Work

Many attempts to visualize OWL ontologies have been presented in the last decade. Surveys can be found in [10,27,32], among others. Most of the works visualize OWL ontologies as graphs, which reflects well the way concepts and relationships are organized in OWL [36]. The graphs are typically rendered in a force-directed, hierarchical, or radial layout, often resulting in appealing visualizations. There are also 3D graph visualizations for OWL ontologies [3,18] as well as approaches that visualize the ontologies as hyperbolic trees [11,15].

However, few visualizations show all concepts modeled in OWL, but most approaches focus on certain aspects of ontologies [35]. While some visualize only the class hierarchy of the ontologies [22,33,37], others consider different types of properties [13,14,45] but do not include property characteristics and other information required to fully understand the information modeled in OWL. Only a small number of works provide comprehensive visualizations for OWL ontologies. Unfortunately, the different ontology elements are partly hard to distinguish in these visualizations. For instance, the tools TGViz [1] and NavigOWL [25] use plain node-link diagrams where all nodes and links look the same except for their color.

SOVA [2] and GrOWL [31] define more elaborate notations using different symbols, colors, and node shapes. However, the resulting visualizations are still comparatively hard to read due to many crossing edges and only minor variations in the visual elements. In addition, both notations use abbreviations and mathematical symbols that make them less intuitive for casual ontology users [35].

There is also a bunch of research on reusing UML class diagrams for the visualization of OWL ontologies [4,8,29]. Precise mappings between OWL and UML class diagrams are, for instance, specified in the Ontology Definition Metamodel (ODM) [41]. However, UML has originally not been designed for the representation of OWL, which results in some conceptual limitations and incompatibilities

[21,29,30]. It can therefore be confusing to illustrate OWL language constructs with UML, especially in teaching and training contexts. Moreover, people not familiar with UML have difficulties interpreting the diagrams correctly, as we found in a comparative user study [38].

A UML-related type of diagram for the visualization of OWL ontologies is used in OntoViz [43]. It groups classes and datatypes in boxes that are linked by object properties. Another proposal has been made with VisioOWL [16], which has been implemented as a template for the diagram editor Microsoft Visio. However, both types of diagrams share the limitation of UML class diagrams that the resulting visualizations are comparatively difficult to read for casual ontology users. Furthermore, the visual distinction of the various OWL elements is again limited, as similar shapes and colors are used for different OWL elements.

A related attempt has been made with Concept Diagrams [24] that consider the logic of OWL in particular. Concept Diagrams aim at a formal representation of ontologies that adequately expresses the OWL semantics. However, they do not provide intuitive OWL visualizations that are also understandable to casual users.

This is different in Graffoo [12] which aims at an easy-to-understand notation for OWL and is therefore closely related to the idea of VOWL. It comes with a comprehensive specification [42] and has been implemented as a GraphML extension for the diagram editor yEd. However, Graffoo is rather related to the idea of UML-based modeling by being based on visual elements commonly known from diagram editors. This makes it less scalable than VOWL when it comes to the visualization of OWL ontologies with many classes, properties, and instances. In addition, and similar to the previous attempts, Graffoo defines a rather general notation that uses similar visual elements for different OWL language constructs. Despite these limitations, Graffoo currently seems to be the most promising alternative to UML, Concept Diagrams, and VOWL when a visual notation for OWL is needed.

3   Summary of VOWL

In general, we designed VOWL as a visual notation that is easily understandable to lay users and that supports the communication between domain experts and ontology experts. We did not attempt to create a direct visual mapping of the OWL *syntax*, but rather focused on the *semantics* of the individual OWL constructs. That is, we were interested in depicting the meaning of the OWL constructs in a way that is easily understandable to lay users, without considering how this meaning is syntactically represented in OWL.

3.1   General Design Principles

For the design of VOWL, we aimed at meeting several design goals that were inspired by the dialog principles defined in ISO 9241-110. This ISO standard "sets forth ergonomic design principles formulated in general terms [...] and provides a framework for applying those principles to the analysis, design and evaluation of interactive systems" [26].

- VOWL was designed to be **self-descriptive** by featuring textual descriptions where necessary or helpful. In the best case, there is no need to consult the VOWL specification or a legend with explanations in order to understand the notation.
- Established patterns were respected by using commonly used shapes and colors (such as arrowheads to indicate direction or gray elements to indicate deprecation), borrowing some visual aspects from notations such as UML. This helps to keep VOWL **conformant with user expectations**.
- We put a focus on the comprehensive specification of the VOWL notation. This complete and unambiguous specification, combined with the self-descriptiveness and conformance with user expectations, makes VOWL **suitable for learning**.
- VOWL is **suitable for its task** of visualizing the semantics of individual and combined OWL elements as well as complete OWL ontologies by defining graphical representations for most OWL language constructs. Users are able to answer a number of questions that are commonly asked when working with ontologies and, in particular, with graph visualizations of ontologies, as we found in user studies we conducted [35,36,38].
- VOWL has been defined in a modular way by making certain labels and colors optional or customizable. Furthermore, we took care to make it work in different interaction contexts, by considering both touch and mouse input for example. The VOWL visualizations scale well with different screen sizes, and can be printed on paper in monochrome, without losing any important information. All of these properties make VOWL **suitable for individualization**.

Accordingly, all of the dialog principles defined in ISO 9241-110 are met by the visual notation except for **controllability** and **error tolerance**, which do not apply to the visual notation itself but rather to its implementation in corresponding tools, as they are summarized in Sect. 4. As already mentioned, the design principles were empirically validated in user studies with different user groups [35,36,38] and a benchmark [36], both focusing the notation itself and its interactive implementations. Besides these primary design principles, we were pursuing some secondary objectives, such as an aesthetically pleasing visualization and a comparatively easy implementation. In addition, we had to make a number of design decisions, which are summarized in the following.

3.2   Concrete Design Decisions

Figure 1 shows a part of version 3.1 of the Geonames ontology [44] visualized with VOWL. The visualization has been generated with the application WebVOWL (cf. Sect. 4).

A key design decision was the basic representation paradigm used to visualize the OWL structure and semantics. The network of classes that can be related

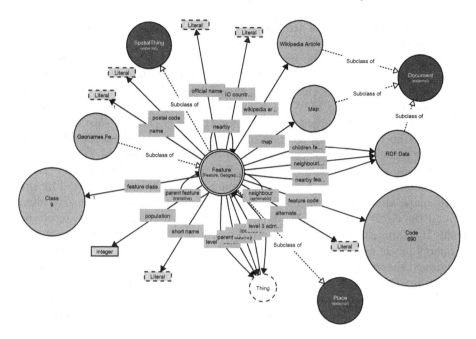

Fig. 1. A part of the Geonames ontology visualized with VOWL.

by inheritance relationships and object properties lends itself to being displayed as a graph. In particular, a node-link-based representation was chosen, as this graph visualization supports the tasks well that are typically relevant in the context of ontology visualization, such as finding an indirect connection between two classes or spotting highly connected classes [28].

Even though indented trees can be more suitable for showing mere hierarchies without multiple inheritance or additional connections by properties, Fu et al. found that node-link visualizations are perceived as "more controllable and intuitive without visual redundancy, especially for ontologies with multiple inheritance" [17]. They are considered particularly "suitable for overviews" and "held [the] attention" better than trees in a comparative user study conducted by Fu et al. [17]. However, it needs to be kept in mind that neither indented trees nor node-link diagrams scale particularly well for very large ontologies.

Each graph element is usually represented exactly once in a node-link visualization of a graph. We decided against this straightforward approach by merging some groups of elements in VOWL that conceptually represent units, such as sets of equivalent classes. Likewise, other elements, such as datatypes and *owl:Thing*, are represented multiple times in the VOWL visualizations. This has the advantage that abstract elements connected with many other ontology elements are prevented from taking central positions in the force-directed graph layout, which is recommended by VOWL and used in the implementations (cf. Sect. 4). At the same time, it allows for shorter edges and fewer edge crossings, both of which enhance the readability of the visualization. As we found in evaluations of VOWL, users are able to correctly interpret the aggregation and multiplication of specific elements [35, 36].

The aggregation and multiplication is one example why VOWL is not a direct mapping of the OWL syntax but of the concepts found in OWL ontologies. Another example are the *disjoint union* constructs which are disassembled into their atomic parts in VOWL, i.e., a union class and pairwise disjoint restrictions between the participants of the union. However, for the sake of a better readability and easier interpretation of the visualization, we are considering to introduce a new element for *disjoint union* in the next version of VOWL.

Two basic shapes, circles and rectangles, were chosen for nodes in VOWL. Class nodes can feature high degrees of connectivity, which is supported by the circle shape that allows higher numbers of inbound arrows to properly align around the circle without overlapping arrowheads. At the same time, datatype nodes that are usually merely connected to one edge have a rectangular shape. Property labels are shown in a rectangle as well, but in contrast to the aforementioned shapes, these rectangles have no frame. Thus, the described elements can be clearly distinguished even in monochrome renditions.

The aforementioned arrowheads are found in inheritance and property relationships. VOWL inheritance edges are reminiscent of UML inheritance and implementation edges, which helps users with the respective prior knowledge [35]. Properties defined in OWL are shown as edges with a label and an arrowhead pointing from the domain to the range. Rather than defining the property as an independent node with two edges and arrowheads to indicate the domain and range, we opted for this graphically simpler visualization as it was preferred by users [38]. Finally, type relationships look similar to properties, but have a special color and label and only appear in cases where a class is an instance of another class.

In general, the size of class nodes can be scaled in VOWL according to the number of individuals that are members of the respective class. It must be pointed out that the exact area covered by circles is difficult to determine for users. Also, a linear mapping between the number of individuals and the radius of the circles may not be helpful, anyway, if the numbers of individuals differ a lot between classes. However, the scaling of class nodes is mainly meant to provide an approximate sense of which classes in an ontology contain slightly or significantly more or less individuals than others at a glance. The exact number of individuals can additionally be displayed inside the class node as text. In this number, individuals are counted for each class they are a member of, also in cases of multi-membership.

Various parts of the visualization were designed to be slightly redundant. For instance, subclass relations are unique in appearance but also carry a textual label. Likewise, deprecated and imported elements have a unique color but also have a descriptive text pointing out their special status. This was done to both improve self-descriptiveness and to minimize the amount of information lost when some features of the visualization are not available. For instance, deprecated classes are still recognizable with impaired color vision or when printed as a monochrome depiction due to the descriptive text. Thus, while colors in

VOWL help to make elements immediately distinguishable, their absence does not imply the loss of crucial information.

3.3 Limitations

Although the graphical representations of the elements described above have many benefits, a few shortcomings still pose open questions in the further development of VOWL. For instance, several reasons beside mere aesthetics support the decision to represent classes with circles, but as labels are often wider than tall, much empty space in class nodes remains. Unless the empty space is used for extra information, the class name can be wrapped, which graphics toolkits are often not able to accomplish within the non-rectangular bounds of a circle. Likewise, IRIs are currently not displayed anywhere in the visualization, as VOWL is directed towards lay users. Those users often do not want to see IRIs, and it is trivial to display them for selected elements in a tooltip and/or next to the visualization.

More generally, VOWL focuses on the visualization of the TBox of small to medium-size ontologies but does not sufficiently support the visualization of very large ontologies and detailed ABox information for the time being. Yet, first attempts to handle larger ontologies by gradually hiding nodes with low degrees of connectivity have been tentatively integrated in the WebVOWL implementation. Future analyses of user needs and suggestions will be required to determine how to integrate more ABox information into VOWL. In particular, this issue must be tackled with respect to ontologies where certain key individuals are as generally used and important as the classes and properties defined in the ontology.

4 Implementations of VOWL

VOWL has been implemented in two different tools that demonstrate its applicability: a plugin for the ontology editor Protégé and a responsive web application. Both tools are released under the MIT license and are publicly available at http://vowl.visualdataweb.org. The OWL ontologies are rendered in a force-directed graph layout according to the VOWL specification. Interaction techniques allow to explore the ontologies and customize their visualizations. While the Protégé plugin is a rather prototypical implementation that does not include all visual elements defined in VOWL, the web application provides a complete implementation of the latest specification of VOWL (which is VOWL 2 at time of writing).

The web application (called WebVOWL) allows users to upload custom ontologies and to interactively explore and adapt the generated VOWL visualizations [34]. It is complemented by a Java-based converter that transforms the OWL ontologies into the required JSON format. The converter parses the ontology representation using the OWL API [23] and outputs a JSON file that is read by WebVOWL. The schema of the JSON file has been designed with regard to VOWL, i.e., its structure differs from common OWL serializations in order to enable an efficient generation of the graph visualization and to ease access at runtime. WebVOWL is easy to use and understand and therefore also appropriate for casual ontology users, as we could confirm in a user study [35].

First attempts to integrate WebVOWL with other approaches have already been started. For instance, WebVOWL is used to visualize OWL fragments in the tool PURO Modeler [9] and to visualize OWL vocabularies in the Linked Open Vocabularies (LOV) service[1]. Ongoing activities related to the implementation of VOWL also concern its integration into WebProtégé[2] as well as the development of a Visio template for VOWL [6].

We have developed a visual query language based on VOWL that addresses the peculiarities of querying Linked Data with SPARQL [20]. Some VOWL elements had to be adapted for this purpose to indicate the variability of the IRIs or values they represent, and to provide for the interactive options that users require to specify their query. For instance, visual elements can act as placeholders in the query language that are not fully specified on a TBox level and for which restrictions can be added by the user. More details on the query language and a prototypical implementation of it are also available at http://vowl.visualdataweb.org.

In related efforts, we have looked into how the VOWL notation could be used to visualize the evolution of ontologies [5] or to represent ontologies extracted from text [7]. Similar to the above query language, we had to incorporate some ABox concepts in the VOWL notation in the latter case. Finally, we developed a benchmark ontology [19] that was used to test the visual scope and completeness of VOWL [36].

5   Conclusions

In this summary paper, we outlined the main design principles and considerations related to the development of a visual notation for OWL. The interested reader is referred to our papers that detail the development of VOWL [35,36,39].

The presented VOWL notation apparently provides only one way to visualize OWL ontologies using node-link diagrams. As OWL is not an inherently visual language, other types of visualizations are also possible and could even be more appropriate in certain cases. For instance, if users are mainly interested in the class hierarchy contained in an OWL ontology, they might prefer a visualization that uses an indented tree or treemap to depict the ontology.

We believe that VOWL is already a comparatively mature proposal to further discuss the visual representation of OWL. However, although version 2 of VOWL already considers a large portion of the OWL language constructs, it is not yet complete, in particular with regard to OWL 2. Our ultimate goal is to turn VOWL into a visual notation that can represent OWL ontologies as completely as possible. Therefore, we recently started work on version 3 of VOWL, with the goal to incorporate additional OWL language constructs.

---

[1] http://lov.okfn.org/dataset/lov.
[2] https://github.com/VisualDataWeb/webprotege.

In addition, we plan to further improve the visual notation, in particular, with regard to the visualization of large OWL ontologies. The current notation does not scale well for large ontologies, as the graph visualization becomes to large, calling for a more compact notation that copes for those cases. Alternatives would be to filter certain elements of the notation or to display only parts of the OWL ontology or abstractions of it. While the WebVOWL implementation already provides some functionality to filter the visualization, this is not yet systematically incorporated into the VOWL specification.

The visual representation of very large ontologies is one of the issues that was discussed at OWLED 2015. Other issues concerned the stability of the visualization and possible alternatives to a force-directed graph layout. We are currently elaborating whether a hierarchical or radial graph layout could also be used together with VOWL. Moreover, we are thinking about ways to integrate further information about individuals defined in OWL ontologies.

At OWLED 2015, we discussed these and other issues and challenges related to the development of a visual notation for OWL. In particular, we gathered feedback on how to further improve VOWL in order to make it even more useful to the community of ontology users.

Acknowledgements. We would like to thank our former and current students David Bold, Vincent Link, and Eduard Marbach for their excellent contributions to the implementation of VOWL and their valuable feedback on the notation.

References

1. Alani, H.: TGVizTab: an ontology visualisation extension for Protégé. In: Proceedings of the 2nd Workshop on Visualizing Information in Knowledge Engineering (VIKE 2004) (2003)
2. Boinski, T., Jaworska, A., Kleczkowski, R., Kunowski, P.: Ontology visualization. In: Proceedings of the 2nd International Conference on Information Technology (ICIT 2010), pp. 17–20. IEEE (2010)
3. Bosca, A., Bonino, D., Pellegrino, P.: OntoSphere: more than a 3D ontology visualization tool. In: Proceedings of the 2nd Italian Semantic Web Workshop (SWAP 2005). CEUR-WS, vol. 166 (2005)
4. Brockmans, S., Volz, R., Eberhart, A., Löffler, P.: Visual modeling of OWL DL ontologies using UML. In: McIlraith, S.A., Plexousakis, D., van Harmelen, F. (eds.) ISWC 2004. LNCS, vol. 3298, pp. 198–213. Springer, Heidelberg (2004)
5. Burch, M., Lohmann, S.: Visualizing the evolution of ontologies: a dynamic graph perspective. In: Proceedings of the International Workshop on Visualizations and User Interfaces for Ontologies and Linked Data (VOILA 2015). CEUR-WS, vol. 1456, pp. 69–76 (2015)
6. Chungoora, T.: Visio template for VOWL (2015). http://ontoweave.com/articles/visio-template-for-vowl/
7. Dasiopoulou, S., Lohmann, S., Codina, J., Wanner, L.: Representing and visualizing text as ontologies: a case from the patent domain. In: Proceedings of the International Workshop on Visualizations and User Interfaces for Ontologies and Linked Data (VOILA 2015). CEUR-WS, vol. 1456, pp. 83–90 (2015)

8. Djurić, D., Gašević, D., Devedžić, V., Damjanović, V.: A UML profile for OWL ontologies. In: Aßmann, U., Akşit, M., Rensink, A. (eds.) MDAFA 2003. LNCS, vol. 3599, pp. 204–219. Springer, Heidelberg (2005)

9. Dudáš, M., Hanzal, T., Svátek, V., Zamazal, O.: OBM2OWL patterns: spotlight on OWL modeling versatility. In: 6th Workshop on Ontology and Semantic Web Patterns (WOP 2015). CEUR-WS, vol. 1461 (2015)

10. Dudáš, M., Zamazal, O., Svátek, V.: Roadmapping and navigating in the ontology visualization landscape. In: Janowicz, K., Schlobach, S., Lambrix, P., Hyvönen, E. (eds.) EKAW 2014. LNCS, vol. 8876, pp. 137–152. Springer, Heidelberg (2014)

11. Eklund, P., Roberts, N., Green, S.: OntoRama: browsing RDF ontologies using a hyperbolic-style browser. In: Proceedings of the 1st International Symposium on Cyber Worlds (CW 2002), pp. 405–411. IEEE (2002)

12. Falco, R., Gangemi, A., Peroni, S., Shotton, D., Vitali, F.: Modelling OWL ontologies with graffoo. In: Presutti, V., Blomqvist, E., Troncy, R., Sack, H., Papadakis, I., Tordai, A. (eds.) ESWC Satellite Events 2014. LNCS, vol. 8798, pp. 320–325. Springer, Heidelberg (2014)

13. Falconer, S.: OntoGraf (2010). http://protegewiki.stanford.edu/wiki/OntoGraf

14. Falconer, S.M., Callendar, C., Storey, M.-A.: A visualization service for the semantic web. In: Cimiano, P., Pinto, H.S. (eds.) EKAW 2010. LNCS, vol. 6317, pp. 554–564. Springer, Heidelberg (2010)

15. Fensel, D., Decker, S., Erdmann, M., Studer, R.: Ontobroker: the very high idea. In: Proceedings of the 11th International Florida Artificial Intelligence Research Society Conference (FLAIRS 1998), pp. 131–135. AAAI Press (1998)

16. Flynn, J.: VisioOWL (2012). http://www.semwebcentral.org/projects/visioowl/

17. Fu, B., Noy, N.F., Storey, M.-A.: Indented tree or graph? a usability study of ontology visualization techniques in the context of class mapping evaluation. In: Alani, H., et al. (eds.) ISWC 2013, Part I. LNCS, vol. 8218, pp. 117–134. Springer, Heidelberg (2013)

18. Guo, S.S., Chan, C.W.: A tool for ontology visualizaiton in 3D graphics: Onto3DViz. In: Proceedings of the 23rd Canadian Conference on Electrical and Computer Engineering (CCECE 2010), pp. 1–4. IEEE (2010)

19. Haag, F., Lohmann, S., Negru, S., Ertl, T.: OntoViBe 2: advancing the ontology visualization benchmark. In: Lambrix, P., Hyvönen, E., Blomqvist, E., Presutti, V., Qi, G., Sattler, U., Ding, Y., Ghidini, C. (eds.) EKWA 2014 Satellite Events. LNCS, vol. 8982, pp. 83–98. Springer, Heidelberg (2015)

20. Haag, F., Lohmann, S., Siek, S., Ertl, T.: QueryVOWL: a visual query notation for linked data. In: Gandon, F., et al. (eds.) ESWC 2015. LNCS, vol. 9341, pp. 387–402. Springer, Heidelberg (2015). doi:10.1007/978-3-319-25639-9_51

21. Hart, L., Emery, P., Colomb, B., Raymond, K., Taraporewalla, S., Chang, D., Ye, Y., Kendall, E., Dutra, M.: OWL full and UML 2.0 compared. Technical report, OMG (2004)

22. Horridge, M.: OWLViz (2010). http://protegewiki.stanford.edu/wiki/OWLViz

23. Horridge, M., Bechhofer, S.: The OWL API: a java API for OWL ontologies. Semant. Web **2**(1), 11–21 (2011)

24. Howse, J., Stapleton, G., Taylor, K., Chapman, P.: Visualizing ontologies: a case study. In: Aroyo, L., Welty, C., Alani, H., Taylor, J., Bernstein, A., Kagal, L., Noy, N., Blomqvist, E. (eds.) ISWC 2011, Part I. LNCS, vol. 7031, pp. 257–272. Springer, Heidelberg (2011)

25. Hussain, A., Latif, K., Rextin, A., Hayat, A., Alam, M.: Scalable visualization of semantic nets using power-law graphs. Appl. Math. Inf. Sci. **8**(1), 355–367 (2014)

26. ISO: ISO 9241–110: Ergonomics of Human-system Interaction - Part 110: Dialogue Principles. ISO (2006)
27. Katifori, A., Halatsis, C., Lepouras, G., Vassilakis, C., Giannopoulou, E.: Ontology visualization methods - a survey. ACM Comput. Surv. **39**(4) (2007)
28. Keller, R., Eckert, C.M., Clarkson, P.J.: Matrices or node-link diagrams: which visual representation is better for visualising connectivity models? Inf. Vis. **5**(1), 62–76 (2006)
29. Kendall, E.F., Bell, R., Burkhart, R., Dutra, M., Wallace, E.K.: Towards a graphical notation for OWL 2. In: Proceedings of the 6th International Workshop on OWL: Experiences and Directions (OWLED 2009). CEUR-WS, vol. 529 (2009)
30. Kiko, K., Atkinson, C.: A detailed comparison of UML and OWL. Technical Report TR-2008-004, University of Mannheim (2005)
31. Krivov, S., Williams, R., Villa, F.: GrOWL: a tool for visualization and editing of OWL ontologies. Web Semant. Sci. Serv. Agents World Wide Web **5**(2), 54–57 (2007)
32. Lanzenberger, M., Sampson, J., Rester, M.: Visualization in ontology tools. In: Proceedings of the International Conference on Complex, Intelligent and Software Intensive Systems (CISIS 2009), pp. 705–711. IEEE (2009)
33. Liebig, T., Noppens, O.: OntoTrack: a semantic approach for ontology authoring. Web Semant. Sci. Serv. Agents World Wide Web **3**(2–3), 116–131 (2005)
34. Lohmann, S., Link, V., Marbach, E., Negru, S.: WebVOWL: web-based visualization of ontologies. In: Lambrix, P., Hyvönen, E., Blomqvist, E., Presutti, V., Qi, G., Sattler, U., Ding, Y., Ghidini, C. (eds.) EKWA 2014 Satellite Events. LNCS, vol. 8982, pp. 154–158. Springer, Heidelberg (2015)
35. Lohmann, S., Negru, S., Haag, F., Ertl, T.: VOWL 2: user-oriented visualization of ontologies. In: Janowicz, K., Schlobach, S., Lambrix, P., Hyvönen, E. (eds.) EKAW 2014. LNCS, vol. 8876, pp. 266–281. Springer, Heidelberg (2014)
36. Lohmann, S., Negru, S., Haag, F., Ertl, T.: Visualizing ontologies with VOWL. Semantic Web (to appear)
37. Motta, E., Mulholland, P., Peroni, S., d'Aquin, M., Gomez-Perez, J.M., Mendez, V., Zablith, F.: A novel approach to visualizing and navigating ontologies. In: Aroyo, L., Welty, C., Alani, H., Taylor, J., Bernstein, A., Kagal, L., Noy, N., Blomqvist, E. (eds.) ISWC 2011, Part I. LNCS, vol. 7031, pp. 470–486. Springer, Heidelberg (2011)
38. Negru, S., Haag, F., Lohmann, S.: Towards a unified visual notation for OWL ontologies: insights from a comparative user study. In: Proceedings of the 9th International Conference on Semantic Systems (I-SEMANTICS 2013), pp. 73–80. ACM (2013)
39. Negru, S., Lohmann, S.: A visual notation for the integrated representation of OWL ontologies. In: Proceedings of the 9th International Conference on Web Information Systems and Technologies (WEBIST 2013), pp. 308–315. SciTePress (2013)
40. Negru, S., Lohmann, S., Haag, F.: VOWL: Visual notation for OWL ontologies (2014). http://purl.org/vowl/
41. OMG: Ontology Definition Metamodel, Version 1.1 (2014). http://www.omg.org/spec/ODM/1.1/
42. Peroni, S.: Graffoo specification (2013). http://www.essepuntato.it/graffoo/specification/current.html
43. Sintek, M.: OntoViz (2007). http://protegewiki.stanford.edu/wiki/OntoViz
44. Vatant, B.: GeoNames Ontology (2012). http://www.geonames.org/ontology/
45. Wachsmann, L.: OWLPropViz (2008). http://protegewiki.stanford.edu/wiki/OWLPropViz

Snap-SPARQL: A Java Framework for Working with SPARQL and OWL

Matthew Horridge[✉] and Mark Musen

Stanford Biomedical Informatics Research Group, Stanford University,
Stanford, CA, USA
matthew.horridge@stanford.edu

Abstract. We present Snap-SPARQL, which is a Java framework for working with SPARQL and OWL. The framework includes a parser, axiom template API, SPARQL algebra implementation, and graphical user interface components for reading, processing and executing SPARQL queries under the SPARQL 1.1 OWL Entailment Regime. While the framework was originally designed to support the implementation of a SPARQL teaching aid in the form of a Protégé plugin, we believe that it is more generally useful and may be of interest to developers and researchers working on SPARQL 1.1 OWL entailment regime implementations and optimisations. The framework is open source and pluggable.

1 Introduction

In March 2013 the World Wide Web Consortium published the SPARQL 1.1 Recommendation—a set of nine documents that specify a query language and protocol for querying and manipulating RDF graphs [2]. Although a point increment, this latest version of SPARQL includes many new language features such as aggregates, sub-queries, a new suite of builtin functions, and path expressions. Besides these many language enhancements, SPARQL 1.1 also includes a sub-specification that describes how SPARQL queries should be evaluated under different *entailment regimes* [3].

An entailment regime is a specification that precisely defines how SPARQL queries should be answered with respect to a given entailment relation. The SPARQL 1.1 specification defines several out-of-the-box entailment regimes, which include the *Simple*, *RDF-Schema*, and *OWL 2 Direct Semantics* entailment regimes. For any given SPARQL query, the set of answers depends upon the entailment regime in question, and the answers for one entailment regime may be different to the answers for a different entailment regime. For example, consider the RDF graph below[1]

```
:PaloAlto :isLocatedInState :California .
:California :isLocatedIn :USA .
:isLocatedInState rdfs:subPropertyOf :isLocatedIn .
:isLocatedIn rdf:type owl:TransitiveProperty .
```

---

[1] Where we omit prefix declarations for the sake of brevity.

© Springer International Publishing Switzerland 2016
V. Tamma et al. (Eds.): OWLED 2015, LNCS 9557, pp. 154–165, 2016.
DOI: 10.1007/978-3-319-33245-1_16

The following query asks for the pairs ?x and ?y where ?x is located in ?y.

```
SELECT ?x ?y
WHERE {
        ?x :isLocatedIn ?y
}
```

Under the *Simple* entailment regime, the set of bindings for ?x and ?y contains the single binding ⟨:California, :USA⟩, while under the *RDFS* entailment regime the set of bindings includes the bindings for the Simple regime plus ⟨:PaloAlto, California:⟩ (entailed in part by the sub-property axiom), and finally, under the *OWL 2 Direct Semantics* entailment regime the set also includes the previous two bindings plus ⟨:PaloAlto, USA:⟩ (entailed in part by the transitive property axiom).

As far as end users are concerned, the practical impact of the SPARQL 1.1 Entailment Regimes specification is that the answers to SPARQL 1.1 queries can potentially include information that follows *implicitly* from the dataset that is being queried. In other words, query answers may now be obtained from *inferred* information, and for OWL users, this means that SPARQL 1.1 can sensibly be used for querying OWL ontologies.

While having a standardised query language for OWL is a huge plus, users also need tools in order to be able to write and execute queries. Furthermore, not only are middleware tools that perform the actual query answering needed, but there is also a need for user-facing tools that assist users in writing queries that are well-formed for both a given entailment regime and a given dataset (set of ontologies). With this in mind, we present a framework called Snap-SPARQL that assists end-users of environments like Protégé in writing SPARQL queries that are well-formed for the OWL entailment regime.

The Snap-SPARQL framework consists of (1) a SPARQL parser, which parses queries that are well formed for the OWL entailment regime and also provides error descriptions that allow meaningful auto-correction/completion to be provided, (2) data-structures for representing basic graph patterns at the level of *axiom templates*, or axioms that may contain variables, (3) an executable implementation of the bulk of the SPARQL 1.1 query algebra that provides the various SPARQL "bells and whistles" beyond basic graph pattern evaluation, (4) a graphical user interface component that contains syntax highlighting and auto-completion, which assists end-users of tools like Protégé in writing SPARQL queries that are well-formed for the OWL Entailment Regime, and (5) a Protégé plugin that uses the aforementioned components to enable Protégé users to query OWL ontologies that are contained in a Protégé workspace.

Finally, the framework is pluggable. It allows reasoners that implement the OWL API [4] reasoner interfaces such as ELK, FaCT++, HermiT, JFact and Pellet to be plugged into it. It also allows basic graph pattern evaluation implementations such as OWL-BGP or the Derivo SPARQL-DL engine[2] to be used for the core graph pattern (axiom template) evaluation operations.

---

[2] http://www.derivo.de/en/resources/sparql-dl-api.html.

```
PREFIX  rdf:  <http://www.w3.org/1999/02/22-rdf-syntax-ns#>
PREFIX  rdfs: <http://www.w3.org/2000/01/rdf-schema#>
PREFIX  foaf: <http://xmlns.com/foaf/0.1/>
PREFIX      : <http://owl.man.ac.uk/2005/07/sssw/people#>

SELECT ?personName ?petName
WHERE {
        ?person rdf:type :person ;
                rdfs:label ?personName ;
                :has_pet ?pet .
        ?pet rdf:type :cat ;
            rdfs:label ?petName

        FILTER (STRSTARTS(?petName, "T"))

        OPTIONAL {
                ?pet foaf:gender ?gender .
        }
}
```

Fig. 1. An example SPARQL query. The query asks for people who have pets that are cats, and lists them along with their names and, if known, gender.

2   Preliminaries

SPARQL is based on Turtle Syntax (Terse RDF Triple Language) [7]. SPARQL queries typically consist of various clauses and blocks, which specify *basic graph patterns* to be matched along with keywords that join, filter and extend the solution sequences to these patterns.

An example SPARQL query, targeted at the *people and pets* tutorial ontology is shown in Fig. 1. The solutions to this query are the people who have cats that have names that begin with the letter "T", along with the name of the cat and, if known, its gender. Curly brackets denote *group graph patterns*, and contiguous triple patterns form *basic graph patterns*. In this query, there are two basic graph patterns split by the OPTIONAL keyword, which performs a *left join* on the solution sequences to the two basic graph patterns. The query prologue contains various *prefix declarations* consisting of a prefix label and a prefix. Various variables are used within the query body, but only three are projected: ?personName ?petName and ?gender.

3   Components and Functionality

The Snap-SPARQL framework contains several components that may be split into programmer-facing and end-user-facing components. We provide a few brief details of each of these components below.

3.1   An Axiom Template API

Axiom templates are essentially OWL axioms that allow variables to be placed in the positions of IRIs and Literals. For example, {ClassAssertion(:person ?x),

AnnotationAssertion(rdfs:label ?x ?y)} is a set containing two axioms templates[3]. The first template is a ClassAssertion axiom that contains the variable ?x in the individual position. Hence bindings for ?x are the entailed instances of :person. The second template is an AnnotationAssertion axiom that contains two variables ?x and ?y in the subject and object position respectively. Intuitively, bindings for ?y are the labels of bindings for ?x. Together, the templates form a query for the labels of the instances of :person.

According to the OWL entailment regime specification, for a query to be considered well-formed it must be possible to "lift" the basic graph patterns in that query into OWL axiom templates. In other words, the basic graph patterns must corresponds to a triple-based serialisation of some OWL axioms.

Axiom templates essentially provide a high-level view of basic graph patterns that makes sense at the level of OWL, and they essentially abstract away from the triple-based syntax of SPARQL. Snap-SPARQL provides an API for working with axiom templates that allows implementors to avoid dealing with triples and the intricacies of the triple-based serialisation of OWL axioms. The axiom template API itself is inspired by the axiom and class expression structures in the OWL API, and it closely follows the OWL functional syntax specification in design.

3.2   A SPARQL Parser

The framework includes a parser that consumes SPARQL syntax and transforms it into high-level axiom templates and other data structures that represent SPARQL queries at an abstract level that is syntax independent. The parser was designed with supporting context-sensitive auto-completion in mind and it forms part of an editing kit that is part of the framework (see Sect. 3.5).

At the time of writing the parser supports most of the SPARQL 1.1 specification in terms of language features. However, some features, such as property path expressions, are not supported. In terms of parsing axiom templates, the parsing of complex class expressions is not *currently* supported, but this is planned as part of future work (see Sect. 4).

Because Snap-SPARQL is designed to handle queries under the OWL entailment regime, the parser will only consume queries that conform to the syntactic restrictions of this regime rather than more general SPARQL queries. In particular, queries must be written so that they can be parsed into well-formed axiom templates, so sets of triples that do not correspond to the triple-based serialisation of OWL axioms, such as {?x rdfs:subClassOf :pete.} (where :pete is an individual) or {?x :hasParent "Mary".} (where :hasParent is an object property), will cause an error. Uses of punning within group graph patterns (between curly braces) will also raise an error, in accordance with the OWL entailment regime specification.

---

[3] For the sake of brevity, we have written these axiom templates using a variant of the OWL Functional Syntax.

Finally, the parser is designed to make it easy for end-users of tools such as Protégé to write SPARQL queries. So for example, the parser does not require all terms appearing in a query to be declared (typed with rdf:type)—if it can determine the type of a term from the underlying ontology then it will do so. Similarly, in some cases the type of variables may unambiguous, for example given {?x rdfs:subClassOf :cat.}, the variable ?x must be of the type owl:Class, i.e. a class variable, and the parser does not require this variable to be typed.

3.3   An Implementation of the SPARQL Algebra

The SPARQL Algebra is a set of operators that together can be used to form *SPARQL Algebra Expressions*. An example of an algebra expression, that corresponds to the concrete SPARQL query shown in Fig. 1, is shown below. An algebra expression, together with data, represents a high-level abstract view of a SPARQL query that is independent from syntactic shortcuts or variations, and syntax-level keywords and punctuation. The algebra is used to define the semantics of SPARQL and it can also be used to derive a canonical procedure for query answering.

```
(Project
    (OrderBy
        (ToList
            (Filter
                (LeftJoin
                    (Bgb
                        ClassAssertion(:person, ?person)
                        AnnotationAssertion(rdfs:label, ?person, ?personName)
                        ObjectPropertyAssertion(:has pet, ?person, ?pet)
                        ClassAssertion(:cat, ?pet)
                        AnnotationAssertion(rdfs:label, ?pet, ?petName)
                    )
                    (Bgp
                        DataPropertyAssertion(foaf:gender ?pet ?gender)
                    )
                )
                BuiltIn(STRSTARTS, ?petName, "T"^^xsd:string)
            )
        )
        OrderCondition(ASC, ?petName)
    )
    ?personName
    ?petName
)
```

Fig. 2. An example of a SPARQL Algebra Expression. This particular algebra expression corresponds to the concrete SPARQL query shown in Fig. 1.

Snap-SPARQL provides an implementation of the SPARQL query algebra for the purposes of query analysis and query plan optimisation, and also to act as a reference implementation for query execution. In the first case, developers of query engines may use the algebra API to generate more optimal query plans. In the second case, developers of implementations that provide basic graph pattern evaluation can simply concentrate on the algorithms for pattern matching and leave algebra operations, such as join, filter, extend, orderby and project to the framework.

3.4 Support for Pluggable Basic Graph Pattern Matching

At the most basic level, answering SPARQL queries involves computing solution-sequences to Basic Graph Patterns (Axiom Templates) and then processing these solution sequences in accordance with the SPARQL algebra mentioned in the previous section. The core operations here are performed by Graph Pattern matching implementations and these are pluggable in Snap-SPARQL. The framework ships with the *Derivo's SPARQL-DL*[4] as the default implementation, but it should also possible to plugin in some other off-the-shelf pattern matcher, such as OWL-BGP [6], with minimal effort. The benefit of this pluggable approach, is that researchers and developers who are interested in supporting the OWL entailment regime, and testing optimisations for query answering under this regime, can focus on axiom template evaluation implementation without worrying about other SPARQL features. In addition to this, Snap-SPARQL enables them to make their implementations available to a wider community, for use in Protégé, without too much extra effort.

3.5 A SPARQL Editor

Writing SPARQL queries can be challenging for users. It requires them: (1) to be fairly well-versed with Turtle syntax in order to construct the basic graph patterns that form the core of any query, (2) to understand the various SPARQL keywords and how these can be used, (3) to correctly setup prefix names and prefixes and then use them consistently in the body of the query, and (4) to use the domain vocabulary in question such that the queries actually make sense. The situation becomes more challenging when basic graph patterns must be well-formed for a given entailment regime such as the OWL entailment regime.

In order to assist users in writing SPARQL queries the framework provides an editor component that can be reused in third party tools. The editor provides the kinds of features that one would expect in a modern development environment such as syntax high-lighting and auto-completion. These features are described in more detail below:

Syntax highlighting of keywords, variables and built-in vocabulary. An example of the syntax highlighting is shown in Fig. 3, where highlighting has been applied to the example query shown in Fig. 1 (note that the highlighting example also includes the addition of comments). Highlighting is applied to keywords, builtin vocabulary, builtin functions and variable names. Furthermore, projected and non-projected variables are distinguished with bold and regular weight fonts.

Auto-suggestion for PREFIX declarations. Dealing with prefixes in SPARQL can be painful. Indeed, this is an issue that related tools have addressed

---

[4] http://www.derivo.de/en/resources/sparql-dl-api.html.

```
PREFIX : <http://owl.man.ac.uk/2005/07/sssw/people#>
PREFIX foaf: <http://xmlns.com/foaf/0.1/>

SELECT ?personName ?petName ?gender
WHERE
{
            # People and their pets
            ?person rdf:type :person ;
                     rdfs:label ?personName ;
                     :has_pet ?pet .

            # We only want the pets that are cats
            ?pet rdf:type :cat ;
                     rdfs:label ?petName

            # Filter pet names that start with T
            FILTER (STRSTARTS(?petName, "T"))

            # Find the gender of the pet - it its present
            OPTIONAL {
                     ?pet foaf:gender ?gender
            }
}
ORDER BY ?petName|
```

Fig. 3. An example of the syntax highlighting used in Snap-SPARQL. Keywords, variables, comments, functions and built-in vocabulary are highlighted. The tool distinguishes between projected and non-projected variables. For example, ?personName and ?petName are highlighted in bold because they are projected into the query result.

(see Sect. 5). Snap-SPARQL provides auto-completion for prefixes based on the IRIs of entities in the signature dataset ontology documents. Well-known prefix names, such as dce:, foaf: or dbo:, are suggested if the corresponding prefixes are present in the underlying dataset. Once prefixes have been declared, the editor will offer completions for terms in the body of the query based on these prefixes rather than offering full IRIs.

Error Highlighting. The editor performs highlighting for different kinds of syntax/semantic errors. Two examples are shown in Figs. 4 and 5, where a double red underline indicates an error. The editor derives error information, both type and position, from the Snap-SPARQL parser. This means that it is capable of detecting several categories of errors, that go beyond SPARQL syntax violation errors that may be detected by a bog-standard SPARQL parser. For example, while both Figs. 4 and 5 show queries that are syntactically correct according to the SPARQL grammar, however they contain other kinds of errors. Specifically, Fig. 4, shows an error where the *class name* :busdriver is not contained in the signature of the dataset ontologies. Figure 5 shows a kind of grammatical error that is only an error under the OWL entailment regime—in this case, the predicate rdfs:subClassOf cannot be used with the variable ?person because this variable will bind to individuals based on its context.

```
SELECT ?personName ?petName
WHERE
{
            # People and their pets
            ?person rdf:type :person ;
                    rdfs:label ?personName ;
                    :has_pet ?pet .

            # This is an error because a class name is expected where we've
            # written :busdriver. However, :busdriver is not a class in the
            # signature of the set of underlying ontologies
            ?person rdf:type :busdriver .
```

Fig. 4. An example of error checking and highlighting. Here, the name :busdriver has been used in a position that should be filled by a class (given the context provided by preceding statements). However, the signature of the underlying set of ontologies does not contain a corresponding term (Color figure online).

```
SELECT ?personName ?petName
WHERE
{
            # People and their pets
            ?person rdf:type :person ;
                    rdfs:label ?personName ;
                    :has_pet ?pet .

            # SPARQL 1.1 does not permit punning within a given group graph
            # pattern. Hence, this is an error because ?person is an individual
            # variable and rdfs:subClassOf expects a class variable as an object.
            ?person rdfs:subClassOf :driver .
```

Fig. 5. Another example of error checking and highlighting. Here, the predicate rdfs:subClassOf has been used in a position that should be filled by an object property, data property or annotation property since :person is an individual variable (given the context provided by preceding statements) (Color figure online).

Context-Sensitive Auto-Completion. In conjunction with the aforementioned error checking, the editor offers context-sensitive auto-completion. Figure 6 shows two examples. On the left-hand side, the variable ?person has not been declared and its type cannot be derived from the context of the query. This means that the editor offers a large choice of possible completions that allow for the variable being a class, property, individual etc. On the-right hand side (after a bit more typing) the parser has a larger context to work with, and it is able to determine that ?person is an individual variable, hence the choice at the cursor is limited to object, data or annotation properties and builtin vocabulary that also applies to individuals such as rdf:type, owl:sameAs and owl:differentFrom. It should be noted that the auto-completion functionality is available in all contexts, including for triple subjects, predicates and objects, as well as key words, function names and punctuation.

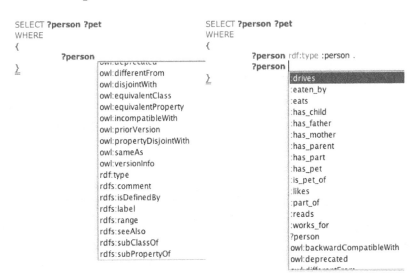

Fig. 6. An example of auto-completion. Auto-completion is provided for keywords, variable names, function names, and terms at all positions.

3.6   A Protégé Plugin

The final Snap-SPARQL component is a Protégé plugin that exposes all of the previously described functionality to end-users of Protégé. In particular, the plugin provides the editing capability described above along with a mechanism to view query results. The plugin is fairly tightly integrated into the Protégé environment, as it uses the ontologies that are loaded into the active Protégé workspace, along with the currently selected reasoner for the purpose of providing inferred information to the basic graph pattern evaluator component of the framework. Finally, the plugin is compatible with Protégé 5, it is open source and may be downloaded from http://github.com/protegeproject/snap-sparql-plugin.

4   Limitations and Future Work

While the Snap-SPARQL framework supports most of the SPARQL 1.1 query language, in particular the features that make sense in the context of the OWL entailment regime there are some features that are not supported. We briefly discuss some of these limitations along with future implementation plans.

SPARQL 1.1 contains *property path expressions* that allow regular-expression-like paths of properties to be matched. However, these are not supported by the Snap-SPARQL framework. While this would be a significant limitation under simple entailment, it is not clear how much of a limitation it actually is under the OWL entailment regime. This is because, one of motivations for property path expressions is that they enable queries to be written whose answers involve some kind of "transitivity" such as {?x rdfs:subClassOf+ ?y} or {?x :partOf+ ?y}.

In these cases, under the OWL entailment regime, transitivity comes "for free" according to the semantics of the language, for example if A is a subclass of B and B is a subclass of C, then A is also a subclass of C. For more complex cases that involve choices e.g. the lack of property path expressions imposes some inconvenience and queries such as {?x rdfs:label | dce:title ?y}, will need to be written by the user, if possible.

SPARQL Update makes it possible to modify graphs in the query dataset. In the case of OWL, this would involve adding and removing axioms to a set of ontologies. While it seems like SPARQL Update for OWL ontologies would be useful, Snap-SPARQL does not currently support this feature.

Specification of specific datasets using the GRAPH keyword is not supported. The framework currently assumes that the query dataset consists of a set ontology documents that are specified by an imports closure in accordance with the OWL semantics. We may support more selective queries using the GRAPH keyword (perhaps restricted to asserted information) in the future.

While Snap-SPARQL supports SPARQL's MINUS keyword, meaning that one solution sequence can be subtracted from another thereby providing a form of negation by failure, it does not currently support NOT EXISTS. We intend to add support for this.

Snap-SPARQL does not currently support complex OWL class expressions. At the moment, queries are essentially limited to SPARQL-DL [10] queries. These kind of queries correspond to mixed ABox/TBox queries over class hierarchies, property hierarchies, disjoint classes, property domain, property range, class assertions, property assertions, same individual, different individuals and annotation assertions. All of this still operates under the OWL entailment regime. Part of the reason for the lack of support for complex class expressions is that these are tricky to write in a triple-based syntax such as turtle, for example ?x SubClassOf hasPart some ?part corresponds to {?x rdfs:subClassOf [rdf:type owl:Restriction; owl:onProperty :hasPart; owl:someValuesFrom ?part]}. We are therefore currently considering whether to support complex class expressions via Terp [9], whereby Manchester Syntax [5] can be embedded into SPARQL queries, or by some form of auto-completion that automatically inserts sets of triple patterns that correspond to well-formed class expressions.

Finally, as part of future work, we are considering supporting explanation of query results using justification based explanation techniques, and we are also considering a web-based version of the editor, in particular for WebProtégé.

5   Related Work

The challenges of writing SPARQL queries using a plain text editor (as provided by many SPARQL end points) are somewhat obvious and have not gone unnoticed by the RDF/SPARQL community. Two tools that are related to the work here are Flint Editor[5] and YASGUI [8], where the latter is based on the

---

[5] http://openuplabs.tso.co.uk/demos/sparqleditor.

former. Both of these tools are Web-based and make it possible to query public SPARQL endpoints. They provide the usual affordances such as syntax highlighting and some form of autocompletion. However, neither of these tools are "entailment regime aware", in particular they are not OWL entailment regime aware. This means that queries are based upon the asserted graph. Furthermore, they do not provide a distinction between domain vocabulary and vocabulary used to encode OWL axioms. They do not perform error checking to the extent that Snap-SPARQL does, which means that they will happily accept nonsensical queries such as `{?x rdf:type :John}`, where John is an individual name, or `{:C :hasPart ?p}`, where `:C` is a class name. Finally, related to the last point, auto-completion is somewhat impoverished when compared to Snap-SPARQL and offers nonsensical suggestions in some cases. For example, attempting to complete `''SELECT ?x WHERE {?x rdfs:subClassOf dbo:Company . ?x dbo:''` offers any kind of predicate that is present in the graph along with language vocabulary— it offers object properties, even though `?x` must be a class variable, and it also offers language vocabulary such as `rdfs:range` or `rdfs:domain`, which does not make sense in this context. On the other hand, both of these tools are fully-fledged general SPARQL 1.1 query tools, they are Web-based, understand commonly used prefixes and offer querying of public SPARQL endpoints.

OWL-BGP [6] is a library for performing basic graph pattern matching under the OWL entailment regime. It also includes an axiom template API and triple consumer for lifting triples into axiom templates before query evaluation takes place. Clients typically do not interact with OWL-BGP directly, as the library itself does not contain a SPARQL algebra implementation. Instead, OWL-BGP interfaces with the Jena RDF framework [1], which provides this functionality. It's possibly the case that OWL-BGP could also be plugged into Snap-SPARQL without too much difficulty.

The Derivo SPARQL-DL library[6] provides basic graph pattern matching under the OWL entailment regime for SPARQL-DL queries. Snap-SPARQL currently uses this library as the default implementation for basic graph pattern matching.

Pellet [11] is an OWL reasoner that supports SPARQL queries over ABoxes. For mixed TBox and ABox queries Pellet falls back to the native Jena RDF SPARQL implementation.

Finally, Stardog[7] is a commercial graph (RDF) database from Complexible that supports OWL reasoning. It supports SPARQL as its native query language, and essentially allows the SPARQL-DL queries to be answered under the OWL entailment regime.

6   Availability

Snap-SPARQL has been developed as part of the Protégé Project. It is open source and freely available at https://github.com/protegeproject/snap-sparql-query.

---

[6] http://www.derivo.de/en/resources/sparql-dl-api.html.
[7] http://stardog.com.

Acknowledgements. This work was supported by Grants GM103316 and R01GM086587 from the National Institute of General Medical Sciences of the United States National Institutes of Health.

References

1. Carroll, J.J., Dickinson, I., Dollin, C., Reynolds, D., Seaborne, A., Wilkinson, K.: Jena: implementing the semantic web recommendations. In: Feldman, S., Uretsky, M., Najork, M., Wills, C. (eds.) Proceedings of the 13th International World Wide Web Conference on Alternate Track Papers & posters, pp. 74–83. New York, NY, USA, ACM, May 2004

2. Glimm, B., Ogbuji, C.: SPARQL 1.1 Entailment Regimes. Technical report, World Wide Web Consortium (2012)

3. Harris, S., Seaborne, A.: SPARQL 1.1 Query Language. Technical report, World Wide Web Consortium (2012)

4. Horridge, M., Bechhofer, S.: The OWL API: a Java API for OWL ontologies. Semant. Web **2**(1), 11–21 (2011)

5. Horridge, M., Patel-Schneider, P.F.: Manchester OWL Syntax for OWL 1.1. In: OWL: Experiences and Directions (OWLED) (2008)

6. Kollia, I., Glimm, B., Horrocks, I.: SPARQL query answering over OWL ontologies. In: Antoniou, G., Grobelnik, M., Simperl, E., Parsia, B., Plexousakis, D., De Leenheer, P., Pan, J. (eds.) ESWC 2011, Part I. LNCS, vol. 6643, pp. 382–396. Springer, Heidelberg (2011)

7. Prud'hommeaux, E., Carothers, G., Beckett, D., Berners-Lee, T.: Turtle terse RDF triple language. Technical report, W3C - World Wide Web Consortium, February 2014

8. Rietveld, L., Hoekstra, R.: YASGUI: not just another SPARQL client. In: Cimiano, P., Fernández, M., Lopez, V., Schlobach, S., Völker, J. (eds.) ESWC 2013. LNCS, vol. 7955, pp. 78–86. Springer, Heidelberg (2013)

9. Sirin, E., Bulka, B., Smith, M.: Terp: Syntax for OWL-friendly SPARQL queries. In: Sirin, E., Clark, K. (eds.) Proceedings of the 7th International Workshop on OWL: Experiences and Directions (OWLED), San Francisco, California, USA, June 21–22. CEUR, vol. 614. CEUR-WS.org (2010)

10. Sirin, E., Parsia, B.: SPARQL-DL: SPARQL query for OWL-DL. In: OWL: Experiences and Directions (OWLED) (2007)

11. Sirin, E., Parsia, B., Grau, B.C., Kalyanpur, A., Katz, Y.: Pellet: a practical OWL-DL reasoner. J. Web Semant. **5**(2), 51–53 (2007)

An Application Ontology to Help Users of a Geo-decision Software Understanding Their Data

Perrine Pittet[(⊠)] and Jérôme Barthélémy

Articque Software, 149 Avenue Général de Gaulle, 37230 Fondettes, France
ppittet@articque.com

Abstract. This paper intends to describe the application ontology of the SaaS version of the decision statistical mapping and geomarketing software Cartes & Données (C & D): CD7Online. Specified in OWL DL, the CD7 ontology was conceived for automation of semantic annotation of CD7Online user data to help users better understand their data and make better selection and representation choices when building maps.

Keywords: OWL DL · Description logics · Application ontology · Ontology development · Semantic annotation · Cartes & données · Geo business software

1 Introduction

Ontologies have been introduced in the Semantic Web research field in the early 2000's to exploit textual documents available on the Web in formalized information [1]. As such, they are sometimes presented as tools for knowledge representation adapted to the Web environment, automatically transforming data into information and information into knowledge [2]. In this paper, we describe an ontology, which was developed to foster users' understanding regarding their data, within a geo business decision SaaS application called CD7Online[1]. This ontology, specified in OWL DL[2], supports the automatized semantic annotation process of user data. In our case the annotation process generates a graph of RDF[3] annotations for each user data workspace, which is stored as a namedgraph in a triplestore. Each namedgraph is automatically queried by an interactive visualization tool, on which users navigate to discover the knowledge behind their data. The rest of the paper is articulated in 4 sections. Section 2 presents the CD7Online project background to expose our motivations for developing a formal application ontology and how this ontology can help users better understand their data. Section 3 describes the CD7 ontology main concepts and justifies their use regarding the task of semantic annotation of user data. Section 4 presents some applications supported by the ontology. Section 5 concludes on feedbacks and future works.

J. Barthélémy—Deceased.

---

[1] CD7Online: https://cdonline.articque.com/.
[2] OWL reference: http://www.w3.org/TR/owl-ref/.
[3] RDF concepts and abstract syntax: http://www.w3.org/TR/2004/REC-rdf-concepts-20040210/.

© Springer International Publishing Switzerland 2016
V. Tamma et al. (Eds.): OWLED 2015, LNCS 9557, pp. 166–173, 2016.
DOI: 10.1007/978-3-319-33245-1_17

2 Project Background

CD7Online is the SaaS application of the 7th version of Cartes & Données[4] (C & D), which is a French commercial decision statistical mapping and geomarketing software, published by Articque[5]. C & D allows users to obtain effective and interoperable maps built on statistical data, without being mapping specialists. As a business decision tool, it is a data analysis and visualization oriented application, which aims at helping people to take decisions via the maps they build upon geo-visualization. C & D has been designed since the very beginning with the aim of being self-explanatory, simple, and highly intuitive for users - ease-of-use being a major requirement. Nevertheless, C & D still relies on the users good knowledge of their data and their ability to choose the relevant analysis and representation tools to build meaningful maps. Also most of C & D users have a punctual use of the software and often do not have enough available time to study and fully exploit the potential of their data. For solving these issues in CD7Online, we decided to provide users the knowledge they require to quickly understand their data and their potential applications. We chose to use an automated semantic annotation process on these data in order to extract and represent this knowledge. Automatized semantic annotation of data is the process of automatically associating relevant metadata to data, so that each data is described by a set of semantic annotations. The main objective is to exploit these annotations to allow users visualizing, via an interactive graph visualization tool, the concepts related to their data and the semantic relations they share. This tool allows them to intuitively navigate in the annotations, compare and select relevant data to build relevant maps (cf. Fig. 1). As CD7Online user data consist in statistical and geographical data tables specified in xml-based files, we adapted a methodology suited to semantic annotation of tables of data proposed in [3]. In [3], an ontology of the food microbiology domain is adapted to support a semantic annotation process. The concepts of this ontology cover the definitions of microbiological symbolic and numerical types, units, value intervals, relations shared by types and the corresponding lexical data, which are used to name them. We similarly developed an ontology describing the knowledge underlying the geographical and statistical data used by CD7Online. Also, because this knowledge strongly depends on the CD7Online application specific uses and processes, this ontology is not a domain ontology as in [3]'s methodology but an application ontology [4]. This however does not alter the efficiency of the semantic annotation process. In fact the methodology has been designed to accept any ontology, in which semantic relations with lexical data can be added, in order to make possible lexical similarity measures. For the development of the ontology, we have followed a simple methodology proposed in [5]. The ontology development and evaluation experience were presented in [6].

The following section focuses on the description of the main concepts of the ontology.

---

[4] C & D website: http://www.articque.com/solutions/cartes-et-donnees/.

[5] Articque website: http://www.articque.com/.

3 CD7 Application Ontology Description

For the purpose of this article, we rely on the ontology definition of [7]. Therefore, we define the CD7 application ontology as a formal explicit description of concepts of the CD7Online data domain, properties of each concept describing various features and attributes of the concepts, and restrictions on properties. The ontology together with the set of individual instances constitute the knowledge base designed for the automatized semantic annotation process. A concept can have subconcepts representing concepts that are more specific than this concept. Properties describe properties of concepts and instances. As we needed to keep the maximum expressiveness while retaining computational completeness and decidability for potential inference purposes, we chose to specify the ontology in the OWL DL language. Note that the CD7 ontology terms are originally written in French. Somehow, to facilitate the reading of its description, we translated terms in English and use description logics [8] in the following part. The CD7 ontology[6] defines two main concepts: *DataComponent* and *CDComponent*. *DataComponent* describes all components related to user data, such as metadata, user data. *CDComponent* describes all components related to the CD7Online specific application processes applicable to user data. All the other concepts fall under these two concepts. Due to lack of space we will focus on the main *DataComponent* underlying concepts, which are used in semantic annotation. Three main concepts are considered: *UserData, Metadata and LexicalData.*

UserData is a *DataComponent* enclosing the three types of data files a CD7Online user can have in a group of his workspace and use within CD7Online, such as statistical data files, basemaps and maps. *UserData* is defined in $\mathcal{SHOIN}(D)$ DL axioms as follows:

$UserData \sqsubseteq DataComponent \sqcap 1\ hasFilename.FileName \sqcap\ \geq 1\ ownedBy.User \sqcap 1\ hasGroup.Group$

$UserData \equiv StatisticalDataFile \sqcup Basemap \sqcup Map$

with *StatisticalDataFile* designating the statistical data files, which contain at least one data table defined by:

$StatisticalDataFile \sqsubseteq UserData \sqcap\ \geq 1\ hasDataTable.DataTable$

with *Basemap* designating basemap files used in maps, containing geographical data of a certain geographical space at a certain geographical level, defined by:

$Basemap \sqsubseteq UserData \sqcap 1\ hasGeographicalSpace.GeographicalSpace \sqcap 1\ hasGeographicalLevel.GeographicalLevel$

with *Map* designating the map project files created by users within CD7Online, which can import basemaps and statistical data columns, defined by:

$Map \sqsubseteq UserData \sqcap\ \geq 0\ hasBasemap.Basemap \sqcap\ \geq 0\ hasDataColumn.DataColumn$

---

[6] CD7 Ontology url: http://support-articque.com/ressources/CD7Ontology.owl.

Metadata is a *DataComponent* designating all the metadata concepts that can be used to describe the underlying knowledge of components of user data or user data themselves in semantic annotations. *Metadata* is defined by:

Metadata ≡ *DataType* ⊔ *GeographicalLevel* ⊔ *GeographicalSpace* ⊔ *DataIndicator* ⊔ *Theme* ⊔ *Date* ⊔ *Unit* ⊔ *WeightedTerm* ⊔ *WeightedWord*

with *DataType* covering three types of data types that can qualify a data column in a data table of a statistical data file: quantitative data, qualitative data and discrete data.

DataType ≡ *QuantitativeData* ⊔ *QualitativeData* ⊔ *DiscreteData* ⊔ *IdData* ⊔ *UnknownDataType*

DataType ⊑ *Metadata* ⊓ $\geq$ 0 *hasDataType⁻.DataColumn*

with *GeographicalLevel* defining all the geographical division levels that can be considered in a statistical data file or a basemap (ex: regional, national level, etc.).

GeographicalLevel ⊑ *Metadata* ⊓ $\geq$ 0 *hasGeographicalLevel⁻.DataTable*

with *DataIndicator* describing all the statistical indicators a data column can be related to (ex: GDP, mortality rate, etc.). Statistical indicators are categorized by themes. Each statistical indicator is associated with at least one weighted term representing the potential composition of weighted lemmas generally used to designate this indicator.

DataIndicator ⊑ *Metadata* ⊓ $\geq$ 0 *hasDataIndicator-.DataColumn* ⊓ $\geq$ 1 *hasTheme.Theme* ⊓ $\geq$ 1 *hasWeightedTerm.WeightedTerm*

Another sort of *DataComponent* is used to support the lexical similarity measures used to determine the statistical indicator related to a data column: *LexicalData*. *LexicalData* designates two lexicons instantiated from two concepts: WeightedTerm and WeightedWord.

WeightedTerm ⊑ *DataComponent* ⊓ $\geq$ 1 *hasWeightedWord.WeightedWord*

with *WeightedWord* defining a lexicon of all the instances of weighted words that can compose weighted terms. A weighted word is described by a text and a weight, which are respectively typed with string and float values.

WeightedWord ⊑ *DataComponent* ⊓ $\geq$ 0 *hasWeightedWord-.WeightedTerm* ⊓ *1 text.xsd: String* ⊓ *1 weight.xsd:float*

Additionally, a set of properties, representing the relations between user data components and metadata (as illustrated above) has been defined. Their domains, ranges and facets have also been formalized (c.f. [6]). Finally, in order to set up the knowledge base for the semantic annotation task, a set of individuals was instantiated from the concepts *WeightedWord, WeightedTerm, DataIndicator, Datatype, Theme, Unit, GeographicalLevel, GeographicalSpace*. These individuals are required for the automatized semantic annotation process. For example, to identify and annotate a data column with a statistical indicator, the process evaluates the lexical similarity of data cells content with instances of WeightedTerm, which are composed of instances of WeightedWord and associated to instances of DataIndicator. Below is illustrated an example of such instantiation:

WeightedWord(w_mortality0.2)
WeightedWord(w_rate1.0)
weight(w_mortality0.2, 0.2)
weight(w_rate1.0, 1.0)
text(w_mortality0.2, « mortality »)
text(w_rate1.0, « rate »)
WeightedTerm(t_mortality_rate)
hasWeightedWord(t_mortality_rate, w_mortality0.2)
hasWeightedWord(t_mortality_rate, w_rate1.0)
DataIndicator(rate)
associatedWeightedTerm(rate, t_mortality_rate)

4   Applications

One of the main features supported by the CD7Online ontology is the semantic annotation of statistical data tables. It implies the identification of the DataIndicator for each data table column.

4.1   Column DataIndicator Identification

The identification of the data indicator related to a data column involves two steps: first the column data type identification, second the data indicators lexical similarity scores according to the column title content.

   The identification of a column data type consists in determining whether its content is quantitative, discrete or qualitative. A set of regular expressions is used to help distinguishing between qualitative numeric values (mostly territorial codes), discrete and true quantitative values. In the ontology, each *DataIndicator* instance is associated to one *Datatype*. Therefore once the data type of a column is identified, the corresponding data indicators lexical similarity scores can be assessed. We adapt here the lexical similarity score definition of [3] (cf. Definition 1). The data indicator, which score is the highest, is then associated to the data column.

Definition 1: Lexical similarity score between column title lemma and weighted terms.

- Let $W = \{w_1 : pw_1;...; w_n : pw_n\}$ and $O = \{o_1 : po_1,..., o_k : po_k\}$, be sets of lemma, with W a set of DataColumn title lemmas w_i and weights pw_i, and O a WeightedTerm instance, with o_i and po_i its respective WeightedWord instances text and weight values.
- Let C be the set of indices pairs (i,j) such as $w_i = o_j$.
- Then the degree of similarity between W and O is:

$$sim_{lex}(w, o) = \frac{\sum_{(i,j)\in C}(p_{w_i} + p_{o_j})}{\sum_{m=1}^{n} p_{w_m} + \sum_{m=1}^{k} p_{o_m}}$$

As recommended in [3], as we do not know which lemma of a column title content W is semantically more important than the others, we automatically associate a default weight of 1.0 to each lemma of the column title considered.

Example: Lexical similarity score of data indicator *gdp* for a column title content *"gdp_per_capita"* composed of two lemma "gdp" and "capita".

DataIndicator(gdp)
hasWeightedTerm(gdp, t_gdp)
WeightedTerm(t_gdp)
hasWeightedWord(t_gdp, w_gdp1.0)
text(w_gdp1.0, "GDP")
weight(w_gdp1.0, 1.0)

If W = ("gdp":1.0; "capita":1.0) and O = ("GDP":1.0) then simlex(W,O) = (1.0 + 1.0)/(1.0 + 1.0 + 1.0) = 2/3.

4.2   Example of Semantic Annotations

A map called "CatchmentArea" created with CD7 is partially described below by a subset of its semantic annotations generated within our system.

```
rdf:type(ns: map298, ns:Map)
ns:nomFichier(ns:map298, "/0/CatchmentArea.cdx")
ns:column(ns:map298, ns:excel55sheet6stores_data_column0)
ns:column(ns:map298, ns:excel55sheet6stores_data_column2)
ns:column(ns:map298, ns:excel55sheet6stores_data_column5)
ns:basemap(ns:map298, ns:basemap15)
```

It uses data columns (c.f. ns:column etc.) from a data table of an excel sheet and a basemap (c.f. ns:basemap). The data column ns:excel55sheet6stores_data_column0 is partially described below.

```
rdf:type(ns:excel55sheet6stores_data, ns:DataTable)
ns:column(ns:excel55sheet6stores_data,ns:excel55sheet6stores_data_colu
mn0)
rdf:type(ns:excel55sheet6stores_data_column0, ns:DataColumn)
ns:dataType(ns:excel55sheet6stores_data_column0,ns:QualitativeData)
ns:indicatorType(ns:excel55sheet6stores_data_column0, ns:store_code)
ns:theme(ns:excel55sheet6stores_data_column0, ns:th_trade)
ns:columnTitle(ns:excel55sheet6stores_data_column0, "store code")
ns:geoLevel(ns:excel55sheet6stores_data, ns:level_town)
```

This data column has been annotated with different metadata: title, the data type of its content (c.f.: ns:dataType), the indicator type (c.f.: ns:indicatorType) and the theme (c.f.: ns:theme) it is related to, the excel sheet (c.f.: ns:column) in which it is contained and the geographical level (c.f.: ns:geolevel) on which its content has been processed.

4.3 Visualization

Semantic annotations can be queried with SPARQL to visualize maps related to a specific theme or indicator, to select basemaps or statistical data files sharing the same geographical levels or time in order to select compatible ones, etc. Figure 1 shows a an example of radial visualization built on dynamic SPARQL querying.

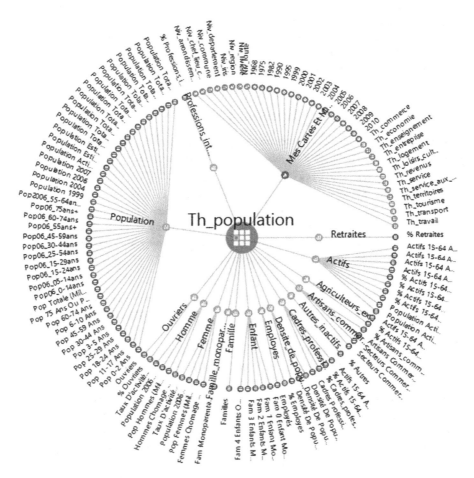

Fig. 1. Visualization of user data annotations within the CD7 graph navigation tool.

5   Conclusions and Future Works

The CD7 application ontology is part of the CD7Online project semantic layer development and supports an automated semantic annotation tool. This tool produces annotations of user data browsable through a graph navigation tool that users can use to better understand their data and build better maps. CD7Online being a commercial software, the development and integration of this layer follows its successive updates. Until now it involved the integration of many semantic tools and technologies. In an industrial project, where deadlines strongly matter, this was a challenge. Hopefully using W3C standards such as OWL clearly helped to reduce development time as many compatible tools for edition, deployment, querying, management and evaluation exist: Protégé, Pellet, Apache Jena-Fuseki, SPARQL, etc. Today, we are working on adding RIF/SPIN rules to provide CD7Online users suggestions of statistical and geographical analysis processes and map representations within a recommender system.

Acknowledgements. This paper is dedicated to the memory of Jérôme Barthélémy, who directed the CD7 project and played a major role in the research work presented here.

References

1. Berners-Lee, T.: Semantic Web Stack (2000)
2. Kaladzavi, G., Diallo, P.F., Lo, M.: OntoSOC: Sociocultural Knowledge Ontology. *arXiv preprint* arXiv:1505.04107 (2015)
3. Hignette, G.: Annotation sémantique floue de tableaux guidée par une ontologie (Doctoral dissertation, AgroParisTech) (2007)
4. Malone, J., Parkinson, H.: Reference and application ontologies. Ontogenesis (2010)
5. Noy, N.F., McGuinness, D.L.: Ontology development 101: a guide to creating your first ontology (2001)
6. Pittet, P., Barthélémy, J.: Experience of formal application ontology development to enhance user understanding in a geo business intelligence saas platform. In: Cuel, R., Young, R. (eds.) FOMI 2015. LNBIP, vol. 225, pp. 51–62. Springer, Heidelberg (2015)
7. Gruber, T.R.: A translation approach to portable ontology specifications. Knowledge Acquisition 5(2), 199–220 (1993)
8. Baader, F., Nutt, W.: Basic description logics. In: Baader, F., Nutt, W. (eds.) Description Logic Handbook, pp. 43–95. Cambridge University Press, Cambridge (2003)

Ontology Engineering: From an Art to a Craft
The Case of the Data Mining Ontologies

Larisa Soldatova[1], Panče Panov[2(✉)], and Sašo Džeroski[2]

[1] Brunel University, London, UK
`larisa.soldatova@brunel.ac.uk`
[2] Jožef Stefan Institute, Jamova Cesta 39, Ljubljana, Slovenia
{`pance.panov,saso.dzeroski`}`@ijs.si`

Abstract. In this paper, we report on our experience and discuss the problems we encountered while designing, implementing and revising a set of ontologies describing the domain of data mining. After giving a short description of the ontologies we have developed, we focus on a set of key issues that we think are important and need to be addressed by the ontology engineering community. These include ontology evaluation, testing, versioning, the use of design patterns, the use of IT portal(s), re-usability, and compatibility. To illustrate the key issues we provide examples that originate from our work on the ontologies for data mining. We conclude the paper with a summary and some suggestions that we believe should be addressed by the ontology engineering research community.

1 Introduction

The progress of ontology engineering for the last two decades has been immense. It is now possible to learn ontologies from text, or construct it following the data-driven approach. Ontologies can be mapped, integrated and re-used. However many steps of ontology engineering remain an art: the design, testing, evaluation. It is still not engineering in a true sense.

For example, in the domain of machine learning, a subfield of computer science that explores the construction of algorithms that learn from and make predictions on data, there are clear procedures how a newly developed algorithm is evaluated, which performance measures should be used for the learning task at hand, and how one can compare with other algorithms defined for the same learning task. Another example is from computer programming, where tools provide complete support and procedures for software testing, versioning, and the use of patterns in the software engineering phase.

In this manuscript, we analyze the state of ontology engineering based on our extensive experience in the development of a number of ontologies, and in particular ontologies for the domain of data mining [1–3], identify the key bottlenecks in its progress and outline a way forward. We illustrate the key issues with examples from our ontologies for the domain of data mining.

© Springer International Publishing Switzerland 2016
V. Tamma et al. (Eds.): OWLED 2015, LNCS 9557, pp. 174–181, 2016.
DOI: 10.1007/978-3-319-33245-1_18

2   Ontologies for the Domain of Data Mining

For the domain of data mining and knowledge discovery, we have developed a modular ontology (named OntoDM) that is composed of three sub-ontologies that can be used together or as a stand-alone product, depending on the use case. This includes: the OntoDM-core ontology that represents core data mining entities [1], the OntoDT ontology that represents datatypes [2], and the OntoDM-KDD ontology that represents the process of knowledge discovery [3].

From the initial version of the ontology [4] to the current ontology releases, we have dealt with variety of problems and design choices largely due to the lack of guidelines for the development of IT ontologies. The simplest design decision in this case was to base our work on the established best practices from the biology and bio-medical domains (such as the OBO Foundry principles[1]), which are still among the most developed domains in regards to ontologies. Other design decisions involved extensive reuse of other ontology resources (classes, relations) from the bio-medical domains using the MIREOT principle [5]. This included the ontology of bio-medical investigations[2] (OBI) [6], information artifact ontology[3] (IAO), and using the basic formal ontology[4] (BFO) [7] as a template at the top level.

In the course of the ontology development, other related ontologies describing the domains of machine learning, data mining and knowledge discovery from various perspectives appeared [8–11]. There was always a problem of how one can compare an ontology to other related domain ontologies and also how to evaluate the constructed ontology sufficiently. Another problem that arose is how to avoid the duplication of the efforts and to reuse classes from other DM ontologies. The reuse has not been straightforward, since the related ontologies were built using different design principles. Some of these issues are discussed further in the following section. In this section, we give a brief summary of the ontologies that we developed and the modeling issues we encountered while trying to represent the domain of data mining.

2.1   Ontology of Core Data Mining Entities

OntoDM-core is developed as an ontology of core data mining entities [1] and it is based on a proposal for a general framework for data mining [12]. It represents the most essential data mining entities in a three-layered ontological structure comprising of a specification, an implementation and an application layer. OntoDM-core provides a representational framework for the description of mining structured data, and it provides taxonomies of datasets, data mining tasks, generalizations, data mining algorithms and constraints, based on the type of data. OntoDM-core is designed to support a wide range of applications/use cases, such as semantic annotation of data mining algorithms, datasets

---

[1] http://obofoundry.org/.

[2] http://obi-ontology.org/.

[3] https://github.com/information-artifact-ontology/IAO/.

[4] http://ifomis.uni-saarland.de/bfo/.

and results; annotation of QSAR studies in the context of drug discovery investigations; and disambiguation of terms in text mining. OntoDM-core is available at http://www.ontodm.com.

2.2 Ontology of Datatypes

OntoDT is developed as a generic ontology for representing the scientific knowledge about datatypes [2]. The ontology is based on an ISO standard for representing datatypes in computer systems [13]. It defines the basic entities, such as datatype, properties of datatypes, specifications, characterizing operations, and a datatype taxonomy. OntoDT was used within the OntoDM-core ontology for constructing taxonomies of datasets, data mining tasks, generalizations and data mining algorithms. Furthermore, OntoDT can be used to annotate and query dataset repositories. In addition, OntoDT can improve the representation of datatypes in the BioXSD exchange format for basic bio-informatics types of data. The generic nature of OntoDT enables it to support a wide range of other applications, especially in combination with other domain specific ontologies: the construction of data mining workflows, annotation of software and algorithms, semantic annotation of scientific articles, etc. The ontology iis available at http://www.ontodt.com.

2.3 Ontology for Representing the Knowledge Discovery Process

OntoDM-KDD is developed as an ontology for representing the knowledge discovery (KD) process [3]. It is based on the Cross Industry Standard Process for Data Mining [14]. OntoDM-KDD defines the most essential entities for describing data mining investigations in the context of knowledge discovery in a two-layered ontological structure. The ontology provides a taxonomy of KD specific actions, processes and specifications of inputs and outputs. OntoDM-KDD supports the annotation of DM investigations in application domains. The ontology has been thoroughly assessed following the best practices in ontology engineering, is fully interoperable with many domain resources and easily extensible. OntoDM-KDD is available at http://www.ontodm.com.

3   The Key Issues

3.1   Evaluation

In our work on ontologies for data mining, we evaluated the produced ontologies using the methodology proposed by Grüniger and Fox [15], which is based on an assessment whether the built ontologies answers the competency questions established in the design phase. In addition, we provided a subjective assessment of how the constructed ontology satisfied the design principles established in the design phase.

While these evaluation approaches are reasonable, they are mainly based on expert opinions: competency questions are created by experts or potential users, and a selection of the design principles is due to experts choice. There are some objective evaluation metrics, e.g. a coverage of the domain, the depth of the hierarchy, a reasoning time. However such metrics do not enable a convincing comparison of ontologies. There are no procedures like for example in machine learning to compare the performance of two algorithms on the same dataset using objective measures.

The development of ontologies is expensive, and therefore it is unlikely to have several ontologies for exactly the same domain to enable an accurate comparison of their performances. However, a comprehensive analyses and evaluation of the key design principles could be performed: what works best for what types of problems/domains: a 4D or a 3D approach; the use of Basic Formal Ontology (BFO) [7] or an ad-hoc upper-level ontology; the use of many or a few relations?

There is a clear need for the research community to come up with a better evaluation approach than those currently available.

3.2   Testing

Testing is an essential part of any engineering project, be it a construction of a bridge or software development. Testing has similarities with evaluation, but it is a different process. On one hand, evaluation aims to identify if a product meets the specified requirement and is closely related with a quality assurance (i.e. a certain level of quality is typically one of the requirements). On the other hand, testing is executing a system in order to identify any errors. Obviously, a system with errors is not a high quality system. Finally, the results of testing are used to improve the system before evaluation.

Standard testing methods for software engineering (e.g. white/grey/black box) can be applied to ontology engineering. However, a development of an ontology has its distinct specificity due to explicit and implicit logical entailments. Unfortunately, there is no methodology available to construct a collection of tests to check if an ontology indeed outputs the expected outputs and does not outputs unexpected ones. Reasoners assist in the detection of logical consistency errors, but they would not detect unexpected outcomes. For example, one can develop a logically consistent pizza ontology where a vegetarian pizza has a meat topping (see the pizza tutorial[5]). Reasoners also do not detect such errors as for example using an `is-a` relation instead of a `part-of` relation. Even if an ontology is error free, and it imports another error free ontology, this does not guarantee that the resulting ontology will be error-free.

The available ontology development tools do not provide sufficient support for the design (manual or automated) and execution of tests. The exception is Tawny-OWL by Lord et al. [16]. The Tawny-OWL library provides a fully-programmatic environment for ontology building; it enables the use of a rich set of tools for ontology development, by recasting development as a form of programming. It is built in Clojure[6] - a modern Lisp dialect, and is backed

---

[5] http://owl.cs.manchester.ac.uk/publications/talks-and-tutorials/
protg-owl-tutorial/.

[6] http://clojure.org/.

by the OWL API. It provides a rich and modern programming tool chain, for versioning, distributed development, build, testing and continuous integration.

The barriers for the development and adoption of testing methodologies for ontology engineering are not only technological, but also sociological. There are expectations that a newly developed ontology should be evaluated before a release or a publication. Unfortunately, it is unusual to report on what set of tests an ontology has been tested. Journals do not require an inclusion of information on testing into papers reporting on ontology development. We as a research community need to change this. We also need to agree on a standard for ontology testing, similar to how it was done for software testing (see the IEEE Standard for Software Unit Testing[7]).

3.3 Versioning

An ontology, like any other any artifact, has a life cycle and it is changing over time. The changes of ontologies may cause interoperability problems. For example, a good practice is not to delete any of the classes, but to deprecate them. But not all developers do that. Ontologies have specificity regarding versioning (see [17]). Changes in ontologies might occur due to various issues and it is important to capture that.

Ontology development tools do not have inbuilt version control of ontology projects. The developers have to use external tools, like Git[8] to compensate for such shortcomings. While it does address the problem of ontology versioning, it is not a seamless process. In terms of versioning support on the level of languages, the OWL language (1.0) provides some built in versioning attributes[9] (e.g., owl:versionInfo, owl:priorVersion), but this is not enough to fully support versioning of ontologies in a systematic way.

3.4 Design Patterns

Mature programming techniques rely on the use of patterns. They speed up the development process and reduce number of errors. Some patterns are available for ontology engineering. Ontology design patterns (ODPs) are ready made modelling solutions for creating and maintaining ontologies. ODPs help in creating rich and rigorous ontologies with less effort [18]. There is a public catalog of ODPs focused on the biological knowledge domain[10]. The OBI (the Ontology for Biomedical Investigations) project developed a pattern (*aka* a template) for the key class assay. Such an approach has significantly speeded up the submission of hundreds of subclasses. Hoehndorf *et al.* provide a prototype to extract relational patterns from OWL ontologies using automated reasoning [19]. However the described efforts are focusing mainly on biomedical areas. There is a need for an easily accessible collection of more generic ontology patterns.

---

[7] http://ieeexplore.ieee.org/xpl/mostRecentIssue.jsp?punumber=2599.

[8] https://git-scm.com/.

[9] http://www.w3.org/2007/OWL/wiki/Ontology_Versions.

[10] https://git-scm.com/html/.

3.5 IT Portal

We deposited our OntoDM and OntoDT ontologies to BioPortal[11]. Currently, it has 450 ontologies and the number is rapidly growing. The portal has an excellent search capabilities and also provides other useful tools, such as: federated querying engine, mapping service, an annotation service, ontology recommender based on an excerpt from a biomedical text or list of keywords and others [20]. However, the portal is supporting biomedical domains, and there is no such a portal for IT ontologies.

The absence of a portal which would provide similar services for IT ontologies contributes to the duplication of the efforts and inhibits the development of the area. For example, we now have several ontologies describing the domains of machine learning, data mining and knowledge discovery that are not interoperable. Having an IT portal with the same functionalities as the BioPortal (or better), would ease the reuse of the ontologies, their discovery and mapping.

3.6 Reausability and Compatibiliy

The most common usage of ontologies remains to be as a controlled vocabulary[12]. However ontologies originally were viewed as 'building blocks' of information systems [21]. We believe that they are under-used in such a capacity and that we will see in the nearest future many interesting use-cases, where ontologies are integral components of complex systems. For example, an ontology LABORS is an integral part of the Robot Scientist system, which is capable of automated discovery of new functional genomics knowledge [22]. The PHenotypic Interpretation of Variants in Exomes (PHIVE) algorithm includes the phenotype manifestations in individuals as well as the signs and symptoms of diseases [23]. This work goes beyond the use of ontologies as controlled vocabularies and exploits hierarchical inheritance properties. It was shown that including phenotype information into the prioritisation of candidate genes leads to an up to 54.1 fold improvement over methods purely based on variant information.

Ontologies as potential information systems 'building blocks' needs to be represented as such to enable their discovery, reuse, integration and functioning as components of complex systems. For example, the service oriented architecture (SOA) approach can be adopted for the description of ontologies as services. SOA is viewed by W3C as a set of components which can be invoked, and whose interface descriptions can be published and discovered. An ontology should be 'wrapped-up' by the specification of what services it can provide, for what domain, what input/output and environment requirements are, provenance, the development stage, quality, etc. A collection of such ontologies would ease the design and implementation of complex information systems.

---

[11] http://bioportal.bioontology.org/.

[12] http://www.w3.org/TR/webont-req/.

4   Discussion and Conclusion

We believe that ontology engineering is still far from reaching its full potential. The developers are struggling with multiple issues outlined above. A better support for the development and evaluation processes would speed up the research in this area. Sophisticated and easy to use development tools can low the barriers for the novice ontology designers and help in reducing errors. Dedicated IT portals would promote best practices in ontology engineering and support ontology reuse and their integration into complex intelligent systems. A mature methodology for the testing and evaluation of constructed ontologies would ensure that the deposited ontologies are of high quality. The research community needs to agree on the standards for ontology testing, description and quality assurance. We are confident that ontology engineering would become a true engineering if these objectives would be achieved.

Acknowledgements. Panče Panov and Sašo Džeroski are supported by The Slovenian Research Agency (Grant P2-0103) and the European Commission (Grants ICT-2013-612944 MAESTRA and KT-2013-604102 HBP).

References

1. Panov, P., Soldatova, L., Džeroski, S.: Ontology of core data mining entities. Data Min. Knowl. Disc. **28**(5–6), 1222–1265 (2014)
2. Panov, P., Soldatova, L.N., Džeroski, S.: Generic ontology of datatypes. Information Sciences. In press (2015)
3. Panov, P., Soldatova, L., Džeroski, S.: OntoDM-KDD: ontology for representing the knowledge discovery process. In: Fürnkranz, J., Hüllermeier, E., Higuchi, T. (eds.) DS 2013. LNCS, vol. 8140, pp. 126–140. Springer, Heidelberg (2013)
4. Panov, P., Džeroski, S., Soldatova, L.N.: OntoDM: an ontology of data mining. In: 2008 IEEE International Conference on Data Mining Workshops. ICDMW 2008, pp. 752–760. IEEE (2008)
5. Courtot, M., Gibson, F., Lister, A.L., Malone, J., Schober, D., Brinkman, R.R., Ruttenberg, A.: Mireot: the minimum information to reference an external ontology term. Appl. Ontol. **6**(1), 23–33 (2011)
6. Brinkman, R.R., Courtot, M., Derom, D., Fostel, J., He, Y., Lord, P.W., Malone, J., Parkinson, H.E., Peters, B., Rocca-Serra, P., et al.: Modeling biomedical experimental processes with OBI. J. Biomed. Semant. **1**(S–1), S7 (2010)
7. Arp, R., Smith, B., Spear, A.D.: Building Ontologies with Basic Formal Ontology. MIT Press, Cambridge (2015)
8. Keet, C.M., Lawrynowicz, A., dAmato, C., Kalousis, A., Nguyen, P., Palma, R., Stevens, R., Hilario, M.: The data mining optimization ontology. In: Web Semantics: Science, Services and Agents on the World Wide Web (2015)
9. Diamantini, C., Potena, D., Storti, E.: A virtual mart for knowledge discovery in databases. Inf. Syst. Front. **15**(3), 447–463 (2013)
10. Vanschoren, J., Soldatova, L.: Exposé: An ontology for data mining experiments. In: International Workshop on Third Generation Data Mining: Towards Service-oriented Knowledge Discovery (SoKD-2010), pp. 31–46 (2010)

11. Esteves, D., Moussallem, D., Baron Neto, C., Soru, T., Usbeck, R., Lehmann, J.: Mex vocabulary: a lightweight interchange format for machine learning experiments. In: SEMANTiCS 2015 (2015)
12. Džeroski, S.: Towards a general framework for data mining. In: Džeroski, S., Struyf, J. (eds.) KDID 2006. LNCS, vol. 4747, pp. 259–300. Springer, Heidelberg (2007)
13. ISO/IEC11404:2007: Information technology - General-Purpose Datatypes (GPD) (2007). URL: http://www.iso.org/iso/catalogue_detail.htm?csnumber=39479
14. Chapman, P., Kerber, R., Clinton, J., Khabaza, T., Reinartz, T., Wirth, R.: The CRISP-DM process model. In: Discussion Paper, March 1999
15. Grüninger, M., Fox, M.S.: Methodology for the design and evaluation of ontologies. In: IJCAI-95 Workshop on Basic Ontological Issues in Knowledge Sharing (1995)
16. Lord, P.: The semantic web takes wing: Programming ontologies with tawny-owl (2013). arXiv preprint arXiv:1303.0213
17. Klein, M.C., Fensel, D.: Ontology versioning on the semantic web. In: SWWS, pp. 75–91 (2001)
18. Aranguren, M.E., Antezana, E., Kuiper, M., Stevens, R.: Ontology design patterns for bio-ontologies: a case study on the cell cycle ontology. BMC Bioinform. 9(Suppl. 5), S1 (2008)
19. Hoehndorf, R., Oellrich, A., Dumontier, M., Kelso, J., Rebholz-Schuhmann, D., Herre, H.: Relations as patterns: bridging the gap between OBO and OWL. BMC Bioinform. 11(1), 441 (2010)
20. Whetzel, P.L., Noy, N.F., Shah, N.H., Alexander, P.R., Nyulas, C., Tudorache, T., Musen, M.A.: Bioportal: enhanced functionality via new web services from the national center for biomedical ontology to access and use ontologies in software applications. Nucleic Acids Res. 39(Suppl. 2), W541–W545 (2011)
21. Mizoguchi, R.: Tutorial on ontological engineering part 1: introduction to ontological engineering. New Gener. Comput. 21(4), 365–384 (2003)
22. King, R.D., Rowland, J., Oliver, S.G., Young, M., Aubrey, W., Byrne, E., Liakata, M., Markham, M., Pir, P., Soldatova, L.N., et al.: The automation of science. Science 324(5923), 85–89 (2009)
23. Robinson, P.N., Köhler, S., Oellrich, A., Wang, K., Mungall, C.J., Lewis, S.E., Washington, N., Bauer, S., Seelow, D., Krawitz, P., et al.: Improved exome prioritization of disease genes through cross-species phenotype comparison. Genome Res. 24(2), 340–348 (2014)

Author Index

Printed in the United States
By Bookmasters